European Association of Archaeologists, 12th Annual Meeting
Cracow, Poland, 19th-24th September 2006

Flint Mining in Prehistoric Europe

Interpreting the archaeological records

Edited by

Pierre Allard
Françoise Bostyn
François Giligny
Jacek Lech

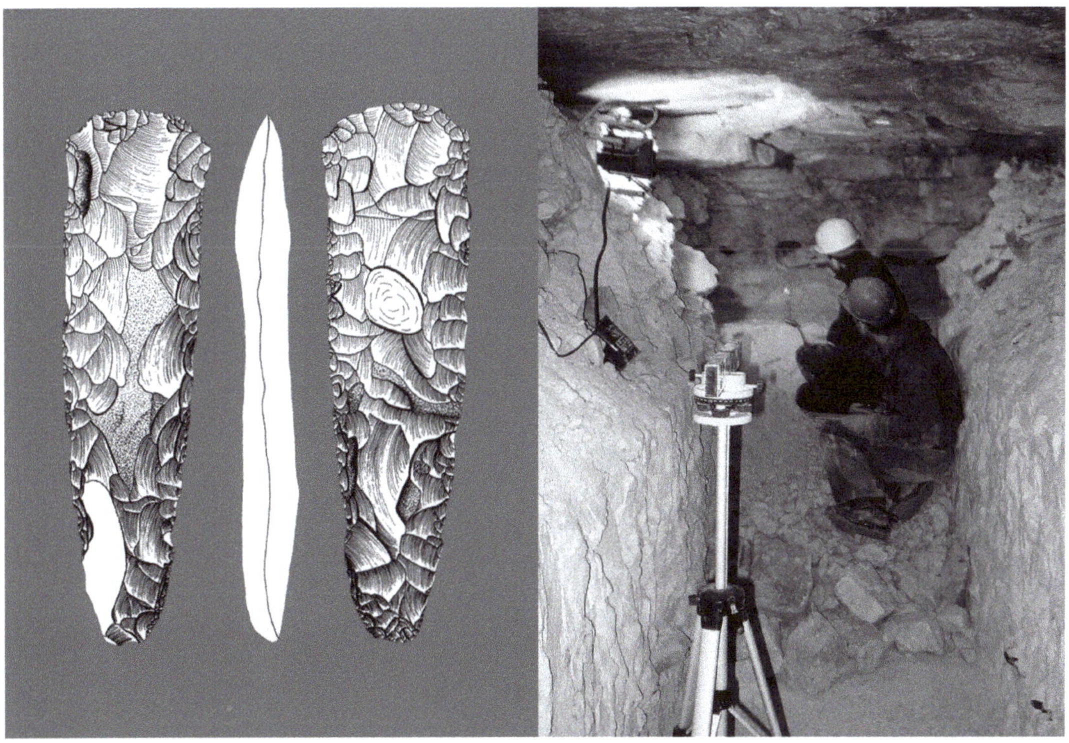

BAR International Series 1891
2008

Published in 2016 by
BAR Publishing, Oxford

BAR International Series 1891

European Association of Archaeologists, 12th Annual Meeting, Cracow, Poland,
19th-24th September 2006
Flint Mining in Prehistoric Europe

ISBN 978 1 4073 0371 0

Editing: Dominique Bossut (Institut National de Recherches Archéologiques Préventives, Direction Nord-Picardie).
Secretary of edition: Françoise Bostyn (Institut National de Recherches Archéologiques Préventives, Direction Nord-Picardie).
The texts have been re-written in part for enhanced clarity and smoothness by Christine Merlié-Young.

BAR Publishing is the trading name of British Archaeological Reports (Oxford) Ltd.
British Archaeological Reports was first incorporated in 1974 to publish the BAR
Series, International and British. In 1992 Hadrian Books Ltd became part of the BAR
group. This volume was originally published by Archaeopress in conjunction with
British Archaeological Reports (Oxford) Ltd / Hadrian Books Ltd, the Series principal
publisher, in 2008. This present volume is published by BAR Publishing, 2016.

Printed in England

BAR
PUBLISHING

BAR titles are available from:

BAR Publishing
122 Banbury Rd, Oxford, OX2 7BP, UK
EMAIL info@barpublishing.com
PHONE +44 (0)1865 310431
FAX +44 (0)1865 316916
www.barpublishing.com

TABLE OF CONTENTS

FLINT EXTRACTION AND PROCESSING FROM SECONDARY FLINT DEPOSITS IN THE NORTH-EAST OF SCOTLAND IN THE NEOLITHIC PERIOD

Alan Saville

Abstract: Secondary flint deposits were of considerable importance to prehistoric people in areas where no such material existed in a primary context. In locations where secondary flint deposits were available close to the surface, multiple shallow pits were often dug over extensive areas. Intensive initial processing of the extracted flint usually took place on site adjacent to the extraction pits, without evidence for tool manufacture. Despite the intensive nature of such extraction, the use and distribution of the resultant flint may be only local and regional. This contribution will consider some of the specific circumstances of such secondary flint exploitation at one particular location, Den of Boddam in Aberdeenshire, north-east Scotland.

Keywords: Den of Boddam, flint, Neolithic, quarrying, Scotland, secondary deposits.

INTRODUCTION

In parts of Europe where there is no flint (or relevant alternatives) available as a fresh raw material in its primary geological context, prehistoric people inevitably sought out other locations where flint or analogous materials were available. Probably the most common sources of such raw material were those locations where flint was exposed at or near the surface by natural processes of erosion, such as on beaches, at cliffs, in river beds, and so on. This was certainly true in Scotland, where there is no flint in a primary geological context. If there ever were flint-bearing Cretaceous chalk deposits on land they have long since eroded away. Flint is locally available, however, in derived and redeposited form, on beaches around some parts of the coast and elsewhere in various glacial, glacio-fluvial and riverine deposits (Marshall 2000; Wickham-Jones and Collins 1978). Flint obtained in this way was used for artefacts throughout the Mesolithic and Neolithic periods and into the Bronze Age.

However, the flint from such deposits was normally only available in small-size pieces, of indifferent quality, and occurred sporadically and somewhat unpredictably, so its exploitation is likely to have been seen as a high-investment / low-return type of activity, perhaps only undertaken expediently in the course of other pursuits. It is, of course, an activity which is difficult to evaluate from the archaeological record. The collection of surface flint pebbles will be very unlikely to leave any direct traces at the collection site, and the collection locations may now be invisible because the occurrences of flint have been exhausted or because of the obscuring effects of erosion, colluviation, peat growth, and other topographic changes.

On the other hand, there are in various parts of Europe geological situations where flint is available in abundant and predictable quantities where it is not in a primary context, but has accumulated as a result of specific processes of erosion affecting the deposits in which it originally formed. These so-called secondary flint deposits (Weisgerber 1987: 131), where they were accessible to prehistoric

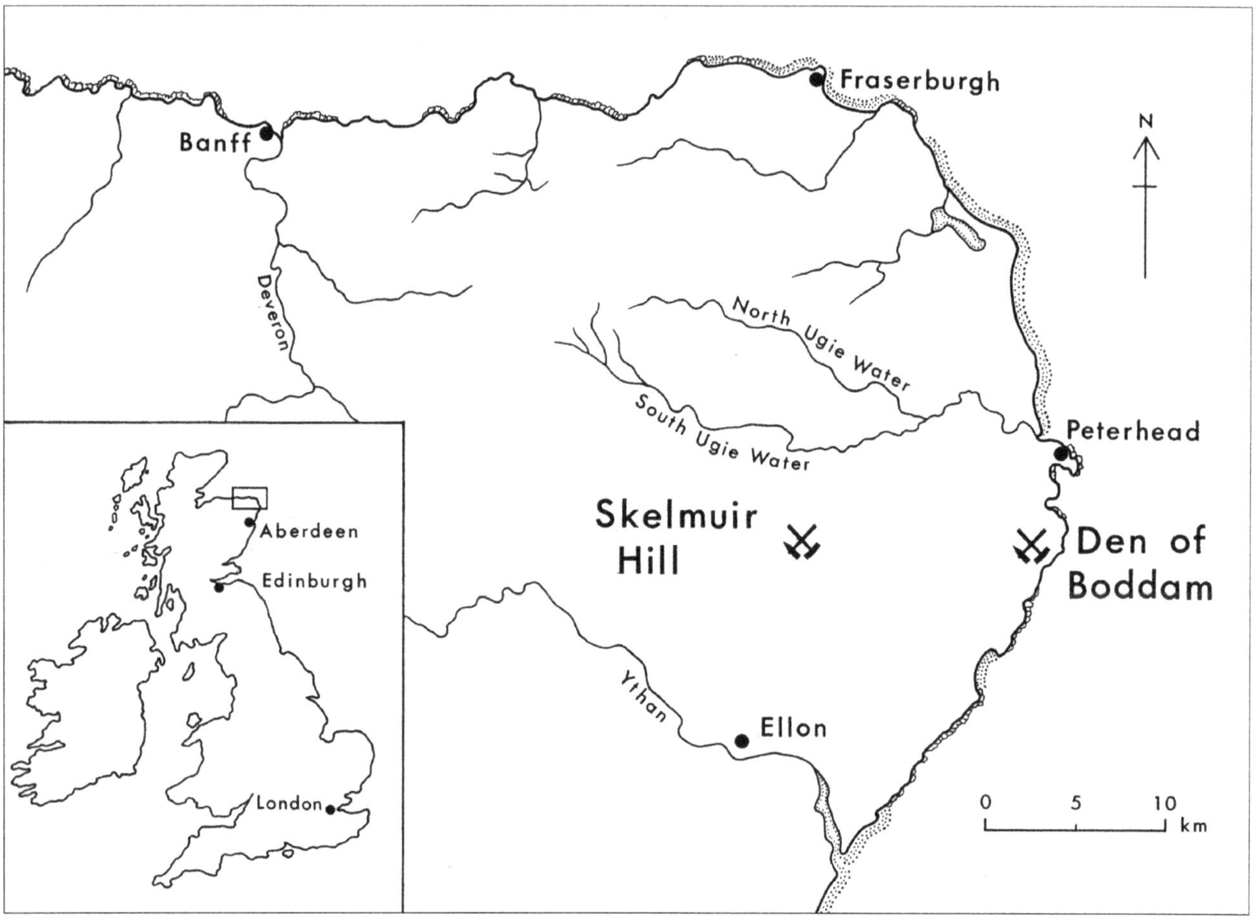

Fig. 1: The flint-rich Buchan Ridge Gravel occurs in a small area south of Peterhead, in Aberdeenshire, north-east Scotland. The two known quarry sites shown, Den of Boddam and Skelmuir Hill (Saville 1995), are approximately at the east and west edges respectively of the occurrence of the Buchan Ridge Gravel deposits.

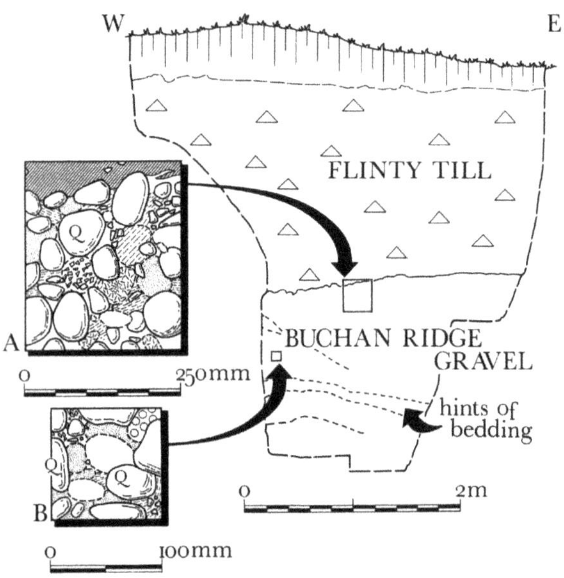

Fig. 2: Den of Boddam. Geological test-pit section, showing the relationship of the Buchan Ridge Gravel to overlying deposits and details of the solid and kaolinized clasts within the Buchan Ridge Gravel. Q = quartzitic cobble; other cobbles and pebbles shown with a solid outline are of flint; those with a dashed outline are decomposed (or 'ghost') clasts of other rock types.

extractive technology, would have formed reliable sources of raw material – albeit not of top quality – for prehistoric people prepared to invest in the labour required to exploit them.

1. DEPOSITS AT DEN OF BODDAM

One such location where a secondary flint source was available and was exploited occurs at Den of Boddam in Aberdeenshire, north-east Scotland (Fig. 1). Here a geological occurrence, known as the Buchan Ridge Gravel Formation, is the result of deposits formed millions of years ago in the Tertiary era, which have become buried inland as a result of subsequent processes of erosion and deposition (Bridgland 2000; Bridgland *et al.* 1997; Merritt *et al.* 2003). As it survives today at Den of Boddam, the variant of these deposits – referred to here as the Buchan Ridge Gravel (and abbreviated to BRG) – consists largely of cobbles and pebbles of flint and quartzite (and similar very robust rocks), thought to be the remains of a fossil marine beach, and where the BRG is overlain by a metre or so of later glacial deposits (Fig. 2). Most other, less resil-

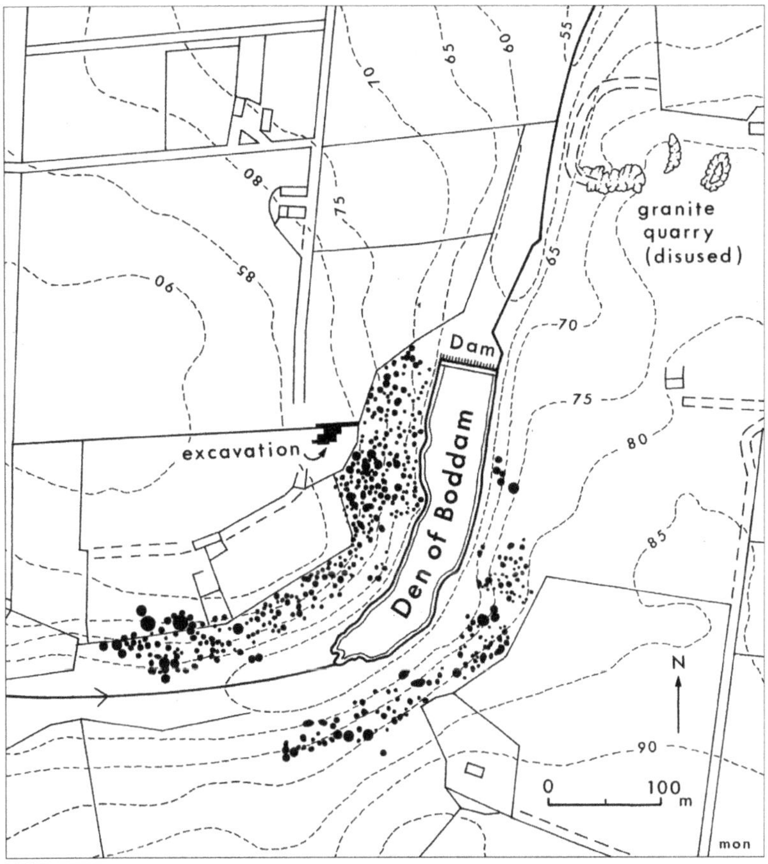

Fig. 3: Den of Boddam location map, showing the position (and limited extent) of the main archaeological excavations in 1992–93. The black dots show the position of all the extraction pits visible on the surface, as plotted by surveyors of the Royal Commission on the Ancient and Historical Monuments of Scotland (RCAHMS 1994: 14-15). The stream running through the Den of Boddam was dammed in recent times to create a reservoir to feed water to a mill further downstream.

ient, lithic components of the BRG have decomposed as a result of deep weathering (Hall 1986) and form the matrix for the surviving flint and other clasts, giving what was in origin an open-framework deposit the character of a matrix-supported one. Beyond this site the BRG survives as a very localized phenomenon, forming the non-continuous capping to the higher parts of undulating low-relief land, extending west from the modern coast over an area about 13 kilometres (eight miles) across.

The existence of the flint-rich deposits at Den of Boddam was identified by prehistoric people, probably following on from their observation of a dense presence of flint cobbles in the stream running through this location, which is a relict glacial meltwater channel (in which the stream is now dammed to create a small reservoir). Prehistoric quarry pits are still plainly visible on the surface as pronounced hollows at Den of Boddam on the steep unploughed slopes of the channel (Fig. 3), making this a unique survival of a Neolithic industrial monument in Scotland.

Samples taken from the same level of the BRG at Den of Boddam as was dug into in prehistory have shown that approximately 68% (by weight) of the deposit is comprised of solid clasts of all kinds and sizes (Fig. 4). Of these, flint clasts sized 50mm or larger (taken as the arbitrary base-line above which size the flint pebbles and cobbles are considered to have been most suitable for prehistoric knapping) constitute by weight approximately 35%, or about a quarter of the total BRG deposit. To give that some context, obtaining 500 flint clasts sized 50mm or larger would require 75.23 kg of the BRG. However, the average maximum dimension of the flint cobbles sized 50mm or larger is only 64mm, and, on the basis of the samples examined, amongst 500 clasts one would expect only 12 cobbles to be 100mm or larger in maximum dimension.

Fig. 4: Den of Boddam. Cobbles and pebbles (washed) extracted from samples of the Buchan Ridge Gravel.

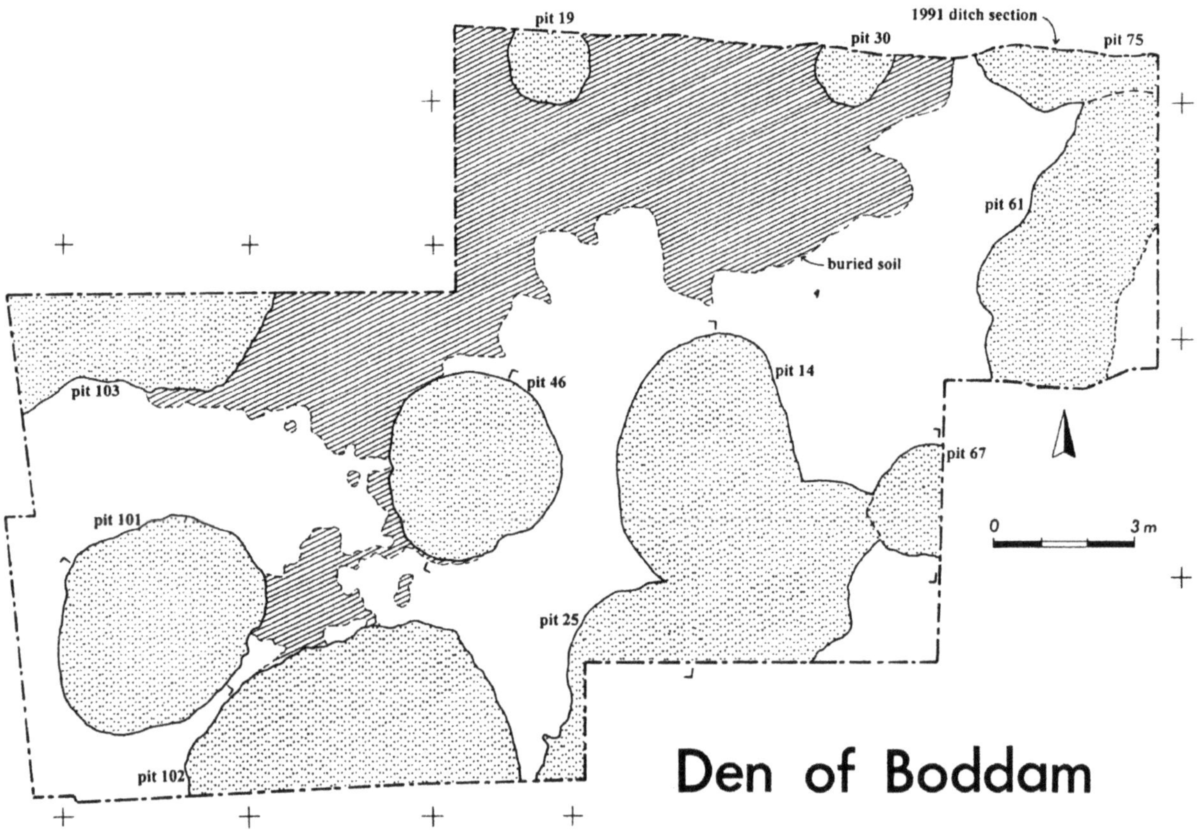

Fig. 5: Den of Boddam. Plan of the pits and other features exposed in 1992–93. Pits 19, 30, 46 and 101 were completely excavated; pits 14 and 67 were half-sectioned. Pits 19 and 30 were shallow features of unknown purpose which did not penetrate as far as the Buchan Ridge Gravel. The buried soil was preserved intermittently between and around pits where modern cultivation had failed to reach the base of the remains of upcast quarry spoil.

Thus if prehistoric people were only interested in large cobbles, extracting BRG would be rather unrewarding labour. All the signs are, however, that smaller pebbles were a viable resource for the Neolithic flint knappers, and that the BRG was a desirable deposit to exploit for flint. There is anecedotal information from the observation of deep sections into the BRG at Den of Boddam and elsewhere in the local area that the availability of larger-sized clasts increases with depth, but the Neolithic miners were unable, given the quarrying technology being employed, to access the deposits very deeply.

All the pebbles and cobbles have a rounded form and chatter-marked cortex surfaces (Fig. 4). Internally the flint is only rarely clear-structured and free of flaws. Coarse inclusions, unsilicified areas, cavities and faults of various kinds are common. Flint from the BRG at Den of Boddam is basically light grey in colour, both internally and externally, reflecting the grey-white colour of the BRG as a whole, which is kaolinized (on account of its original-high granite content) following deep weathering, and this process has acted to effectively 'bleach' the exterior and interior of the flint cobbles. Flint is also present within the metre-thick glacial gravel which overlies the BRG, and comprises the same type of rounded, chatter-marked cobbles, but these are distinctively different in appearance, having a dark brown to black cortex and a grey-brown to brown interior. In those areas investigated by excavation at Den of Boddam it is clear that for the most part this flint from the glacial gravel was ignored, since debitage which is brown in colour represents only a tiny proportion of the total recovered.

Neither in size nor in quality, therefore, is the Den of Boddam flint suitable for the manufacture of axeheads or other large implements, only for making smaller flake implements such as scrapers, knives and arrowheads. The relatively few flint axeheads found in north-east Scotland are virtually all made of flint imported from outside the region (Saville 1994, 1999).

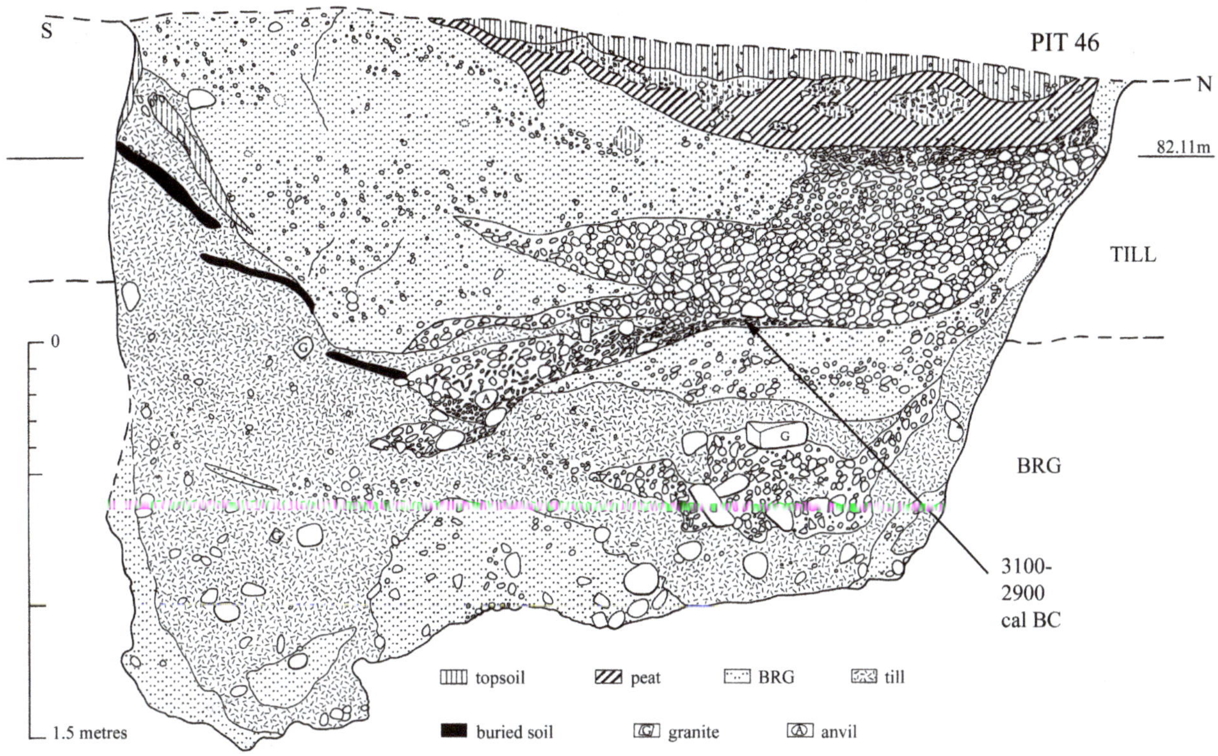

Fig. 6: Den of Boddam. Pit 46 north-south section drawing, showing the position of the birch charcoal sample which produced a radiocarbon age of 3100-2900 cal BC (OxA-13103; table 1). The sample was located at the base of a substantial infill deposit of flint-knapping debris. The buried soil inclusions within the fill on the south side of the pit indicate an area where the pit edge has collapsed over an undercut into the Buchan Ridge Gravel.

2. QUARRYING

Although the existence of the flint-bearing BRG deposits at Den of Boddam must have been recognized since the Later Mesolithic period, from when there exist the first signs of human inhabitation of the local area, it is not until a developed stage of the Neolithic that invasive exploitation by quarrying takes place. The extraction method employed involved the repeated excavation of roughly circular, cylindrical pits through the overlying topsoil and glacial gravel and down into the BRG (Fig. 5). The deepest pits (amongst the very small sample of pits investigated thus far) reached just over 4m below the surface, and even this must have been high-risk for those extracting the flints, because the BRG and glacial gravel deposits are inherently unstable. At Den of Boddam the absolute thickness of the BRG is unproven, but it is known to continue with an equal density of flint for many metres below the depth of the deepest prehistoric pits. In virtually every instance the pits examined archaeologically have shown clear signs of edge collapse (Fig. 6). This is unsurprising given the instability of the deposits, especially as there are indications in many cases at Den of Boddam that when first dug the pits were bell-shaped (that is they expanded at the base, presumably to maximize the yield of flint per pit and because it was the flint from the BRG which was required, and not that from the overlying glacial gravel). There is an inevitability, when dealing with pits in unstable deposits, that the excavated shapes of the pits will generally bear only a partial relationship to their original form (Budziszewski 1997).

The prehistoric exploitation of the BRG was thus a matter of simple extractive technology, but extraction which would have required careful management of spoil if extraction capacity was to be controlled and in any sense maximized. In some cases the fact that there are gaps between what appear to have been perfectly productive pits may indicate the position of former spoil heaps; in other cases the pits overlap or were contiguous (Fig. 5).

The scale on which extraction took place at Den of Boddam was considerable. There is still the surface evidence for at least 458 pits (Fig. 3). Following archaeological investigations (Bridgland and Saville 2000; Saville 1995), it is estimated that perhaps as many as 1000 pits were originally dug at this location over an area of almost 12 hectares (30 acres). In crude spatial terms this would make the extraction at Den of Boddam more extensive than at any of the English flint mines on the Chalk, except Easton Down, Wiltshire (Barber *et al.* 1999: 58), though this is not comparing like with like. Grimes Graves in Norfolk may only have somewhat over 400 shafts across an area of eight hectares, but of course the shafts there relate to intensive underground mining and the amount of high-quality usable flint extracted at Grimes Graves must exceed that at Den of Boddam by an enormous factor. (Note that the extent of the mined area at Grimes Graves is disputed; for example, Lech and Longworth [2000: 70] claim it to be 25 hectares.)

Very approximate calculations can be made to estimate the yield from the Den of Boddam quarry pits, on the basis of the very small sample of the site which has been archaeologically investigated. Ignoring the superficial soil horizon and the glacial gravel which overlies the BRG, the area of BRG extracted from a quarry pit would be in the region, rather conservatively estimated, of six cubic metres. On the data provided by the BRG samples discussed above, this amount of BRG would be likely to contain some 12,720 flint cobbles sized 50mm or larger. Unfortunately the excavations at Den of Boddam have provided no data on the number or percentage of the numerous unstruck flint cobbles found among the backfill deposits in the pits, but even if one were to assume that only half of the number of cobbles extracted were subject to, at the minimum, some testing, and that only half the number of those went on to be knapped to some extent, this would still be in the order of 3000+ cobbles sized 50mm or over per quarry pit. Compared to the serendipity of finding flint cobbles from local beaches and chance exposures, the resource available from quarrying BRG must have seemed well worth while.

Dating the quarrying activity at Den of Boddam has proved difficult, both in relative terms in the absence of diagnostic lithic tool-types or other material culture, such as pottery, and in absolute terms because of the lack of organic materials for radiocarbon dating (the acidic deposits are not conducive to the preservation of items of bone, antler, or wood). Charcoal is present on the site but only in a very few instances during excavation were samples of charcoal recovered from useful secure contexts where the amount or size of the charcoal precluded the possibility of confusion resulting from post-depositional movement. However, two secure samples, both of birch charcoal, taken from infill horizons within quarry pits, give two termini ante quos of c. 3000 cal BC for the quarrying activity in one part of the site (see Table 1). The position of one of these samples, at the base of an infill horizon of knapping debris, is probably at the surface at what was a relatively swift initial collapse and infill of the lower part of the pit (Fig. 6), implying that the date for the extraction pit might not be very much earlier than the radiocarbon date. Some confirmation for this might be shown by radiocarbon dates from an area of buried soil at Den of Boddam

Lab code and no.	Site coding and sample no.	Sample type	Context	$\delta^{13}C$ ‰	^{14}C years BP	Cal date BC OxCal v3.10 @ 95.4%
GU-3438	DB'91/9002	buried soil / humic fraction	Area 1, 'ditch' section	-29.3	4580 ± 60	3520–3090
GU-3439	DB'91/9002	buried soil / humin fraction	Area 1, 'ditch' section	-28.9	4530 ± 50	3370–3030
OxA-13102	DB'91/357	birch charcoal	'ditch' Area A west; pit infill within knapping debris	-26.9	4372 ± 35	3090–2900
OxA-13103	DB'93/323	birch charcoal	Area 3, Pit 46, base of context 48 (knapping debris)	-25.1	4387 ± 34	3100–2900

Table 1: Radiocarbon dates from Den of Boddam.

sealed beneath dumps of upcast material from the quarry pits, which gave a result in the second half of the 4th millennium cal BC (Table 1).

There are no chronological indicators for the length of time over which quarrying took place at Den of Boddam, so although from the number of pits we have an idea of the large scale on which quarrying took place, we have no idea of the intensity of this activity. If the real total of pits at this location is 1000, then 5 pits dug per year would represent a 200-year span, 10 pits dug per year a 100-year span, 20 pits a 50-year span, and so on (and this assumes that pits were dug every year, which is of course another unknowable factor). Equally, there is no archaeological evidence which can help indicate the number of people who might have been involved in the processes of quarrying and knapping at Den of Boddam, and for how much time per year they devoted to these tasks. So at the moment there is no factual basis for assessing which year span might be appropriate, but it might be guessed that 20 pits per year (yielding 250,000+ cobbles) would be beyond the raw material requirement, if not beyond the manpower resources deployable.

3. PROCESSING

In the area investigated archaeologically at Den of Boddam the flint extracted from the quarry pits was processed by primary knapping immediately adjacent to the pits. Cobbles and pebbles were tested and either immediately discarded if considered inappropriate for additional working, or they were subjected to further knapping, occasionally to the extent of producing exhausted cores. Some cobbles were opened by free-hand knapping using a hammerstone, with the cobble being held in the hand while struck, others were opened by the anvil technique. In the latter case the flint cobble was placed and held on a cobble anvilstone (usually a quartzitic cobble) before being hit with a stone hammer. The deciding factors involved in the choice of opening technique are unknown; logically it might be expected that the anvil technique would be employed for the more rounded cobbles and the freehand knapping for the more elongated or sub-angular cobbles which already proffered a usable platform or on which one could easily be created. On the other hand there could have been personal preference for one technique over another on the part of individual knappers.

Experience of digging through the BRG while excavating showed that, particularly when conditions were wet, the matrix became very sticky and clinging, adhering to the clasts to the extent that it was often difficult to discriminate between flint and non-flint, let alone between which flint cobble might be more amenable to hand knapping than another. It might of course be the case that in such conditions (which would be dangerous) the Neolithic quarriers might either not work at all, or they might stockpile the clasts to allow them to 'weather-off' naturally. The fact that quartzitic cobbles occasionally show signs of knapping, and that flint cobbles were occasionally used as anvils, suggests that confusion did sometimes exist.

Soft-hammers of wood or antler would not survive in the soil conditions at Den of Boddam, but in fact the debitage left on site shows virtually no indication of anything other than hard-hammer flaking. The hammers are predominantly cobbles of quartzite and analogous hard stones and only rarely of flint. They vary in size and the extent to which they have been used, some being abandoned while still retaining the hammer surface(s) intact, others after clearly having spalled and fractured during use (Fig. 7).

Some hammerstones have been used as anvilstones (or vice versa). Anvilstones, almost always non-flint cobbles, also vary widely in size and extent of usage. The pits which result from knapping can be single, minimal, and hardly visible, perhaps indicative of use for opening only one flint cobble, or multiple and deep from repeated usage, and these stones frequently broke while in use (Fig. 8). The anvilstones are common – over 240 were recovered from the recent excavations at Den of Boddam – but at this site their use appears primarily to have been in the testing and opening of cobbles, rather more than for bipolar core knapping as such, since the evidence for bipolar reduction amongst the debitage is limited. Conventional platform cores (Fig. 9), and sub-discoidal, sub-Levallois cores are the two most common methods of flake production. The later type often approximates to the 'tortoise core' form, whereby a cortical, often primary, flake from the edge of a cobble (whether anvil struck or not), is trimmed around part of the periphery by removals struck from the ventral, bulbar surface, then a flake (or sometimes more than one) is removed across the ventral surface, having a faceted platform determined by the previous peripheral flaking (cf. Saville 2006).

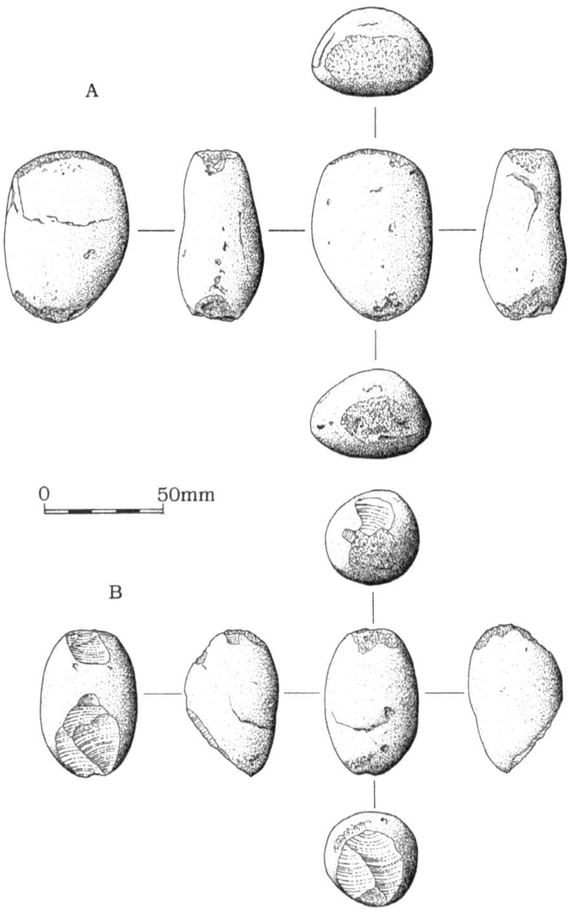

Fig. 7: Den of Boddam. Hammerstones.

Fig. 8: Den of Boddam. Anvilstones.

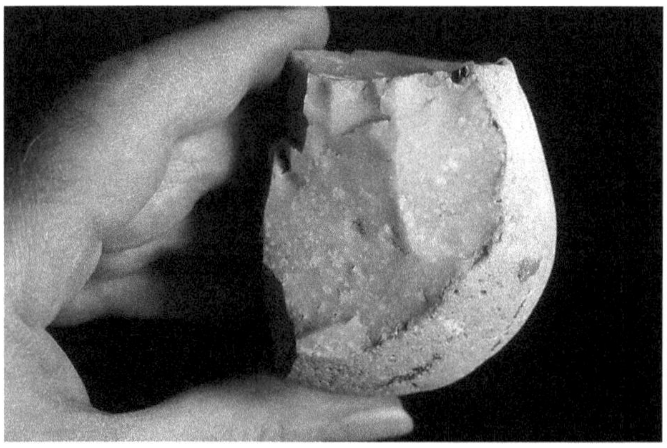

Fig. 9: Den of Boddam. Platform core.

Archaeological investigations at Den of Boddam have resulted in the recovery of hundreds of thousands of pieces of flint debitage, but have not yet discovered any finished (or even roughout versions) of implements which might be seen as the intended product of this large-scale enterprise. There are occasional ad hoc tool-forms such as minimally retouched scrapers and irregular pieces with areas of secondary retouch, but no recurrent types of formal implements. Probably the most parsimonious explanation for the absence of finished implements or roughouts / preforms at this quarry site is that the intended product on-site was the flake blank, and that further processing of the blanks took place off-site, or at least in a different part of the site from any which has yet been investigated. This would explain both the absence of the implements themselves, and also the absence of successfully-produced, reasonably well-sized and well-proportioned flakes amongst the residual debitage. The most likely end product from those flakes is considered to be arrowheads (Saville 2005), but for the moment this must remain a suggestion rather than a fact.

Specific fieldwork and the study of museum collections to track the Neolithic distribution of Buchan Ridge Gravel products has not taken place in any rigorous fashion, but casual observation would suggest that this flint was not dispersed far beyond the north-east of Scotland. The distribution would appear to be an easterly one, from down towards Dundee in the south to around Inverness to the west (i.e. a distance of perhaps 130 km [80 miles] in each direction beyond the in situ Buchan Ridge Gravel deposits). It is not possible, however, to discriminate between flint which may have been obtained directly from extraction pits at Den of Boddam or elsewhere, and flint which has been derived from the Buchan Ridge Gravel and obtained from natural exposures or superficial deposits, around the coast, in river beds, or elsewhere. This unsatisfactory situation is of course compounded by current ignorance of the actual intended outcome of the extraction activity in terms of implement types. It is hoped that ongoing studies will be able to throw more light on the matter.

Alan Saville
Archaeology Department,
National Museums Scotland
Chambers Street, Edinburgh EH1 1JF, Scotland, UK
e-mail: a.saville@nms.ac.uk

REFERENCES

Barber, M., Field, D. and Topping, P., 1999. The Neolithic Flint Mines of England. Swindon: English Heritage.

Bridgland, D.R., 2000. Discussion: the characteristics, variation and likely origin of the Buchan Ridge Gravel. In J.W. Merritt, E.R. Connell and D.R. Bridgland (eds.), The Quaternary of the Banffshire Coast and Buchan: Field Guide, London: Quaternary Research Association, 139-143.

Bridgland, D.R. and Saville, A. 2000. Den of Boddam. In J.W. Merritt, E.R. Connell and D.R. Bridgland (eds.), The Quaternary of the Banffshire Coast & Buchan: Field Guide, London: Quaternary Research Association, 102-115.

Bridgland, D.R., Saville, A., and Sinclair, J.M., 1997. New evidence for the origin of the Buchan Ridge Gravel, Aberdeenshire. Scottish Journal of Geology 33, 43-50.

Budziszewski, J., 1997. Mine 1/4 of the 'Za Garncarzami' mining field in Ozarow (Central Poland): remarks on methods of studying shallow flint mines. In A. Ramos-Millan and M.A. Bustillo (eds.), Siliceous Rocks and Culture, Granada: Universidad de Granada, 151-162.

Hall, A.M., 1986. Deep weathering patterns in north-east Scotland and their geomorphological significance. Zeitschrift für Geomorphologie 30, 407-422.

Lech, J. and Longworth, I., 2000. Kopalnia krzemienia Grimes Graves w swietle nowych badan. Przeglad Archeologiczny 48, 19-73.

Marshall, G., 2000. The distribution of beach pebble flint in western Scotland with reference to raw material use during the Mesolithic. In S. Mithen (ed), Hunter-gatherer Landscape Archaeology: the Southern Hebrides Mesolithic Project 1988-98, Vol.1, Cambridge: McDonald Institute, 75-77.

Merritt, J.W., Auton, C.A., Connell, E.R., Hall, A.M. and Peacock, J.D., 2003. Cainozoic Geology and Landscape Evolution of North-East Scotland. Edinburgh: British Geological Survey.

RCAHMS, 1994. Monuments on Record: Annual Review 1993-4. Edinburgh: Royal Commission on the Ancient and Historical Monuments of Scotland.

Saville, A., 1994. Exploitation of lithic resources for stone tools in earlier prehistoric Scotland. In N. Ashton and A. David (eds.), Stories in Stone, London: Lithic Studies Society, Occasional Paper 4, 57-70.

Saville, A., 1995. Prehistoric exploitation of flint from the Buchan Ridge Gravels, Grampian Region, north-east Scotland. Archaeologia Polona 33, 353-368.

Saville, A., 1999. An exceptional polished flint axe-head from Bolshan Hill, near Montrose, Angus. Tayside and Fife Archaeological Journal 5, 1-6.

Saville, A., 2005. Prehistoric quarrying of a secondary flint source: evidence from north-east Scotland. In P. Topping and M. Lynott (eds.), The Cultural Landscape of Prehistoric Mines, Oxford: Oxbow Books, 1-13.

Saville, A., 2006. Flint technology and production associated with extraction sites in north-east Scotland. In G. Körlin and G. Weisgerber (eds.), Stone Age – Mining Age, Bochum: Deutsches Bergbau-Museum (= Der Anschnitt 19), 449-454.

Weisgerber, G., 1987. The technological relationship between flint mining and early copper mining. In G. de G. Sieveking and M.H. Newcomer (eds.), The Human Uses of Flint, Cambridge: Cambridge University Press, 131-135.

Wickham-Jones, C.R. and Collins, G.H., 1978. The sources of flint and chert in northern Britain. Proceedings of the Society of Antiquaries of Scotland 109 (1977-78), 7-21.

ACKNOWLEDGEMENTS

I am grateful to Françoise Bostyn for the initial invitation to speak in the flint mining in Prehistoric Europe session at the EAA Conference in Krakow in 2006, and for her encouragement and patience while awaiting the written version for this volume. Figs 1 and 3 were drawn by Marion O'Neil, Figs 2, 5 and 7 by Alan Braby, and Fig.6 and the photographs are by the author. I wish to thank the Royal Commission on the Ancient and Historical Monuments of Scotland for permission to incorporate the results of their survey work in Fig.3, and Craig Angus for help with images.

FLINT WORKING AT THE EARLY LINEARBANDKERAMIK SETTLEMENT OF GELEEN-JANSKAMPERVELD

(PROVINCE OF LIMBURG, THE NETHERLANDS)

Marjorie E. Th. de Grooth

Abstract: The main objective of this study was to investigate the way flint working was organized at the Early LBK settlement of Geleen-Janskamperveld, thus getting an insight into the procurement strategies and the technological choices made at the beginning of Bandkeramik settlement west of the river Rhine. A new evaluation of raw material characteristics made it possible to determine the (probable) extraction site, located around 22 km south of the settlement. In the Banholtergrub, a dry valley 5 km to the east of the Rijckholt flint mines, nodules originating from the upper Cretaceous Lanaye chalk layer were extracted from residual loams. Intensive processing took place in the settlement; surplus material was distributed in the shape of partially reduced blade cores and blades. A comparison with the contemporary settlements at Elsloo (Graetheide plateau) and Langweiler 8 (Aldenhovener Platte) made it clear that different exchange mechanisms prevailed, so that Langweiler 8 may have received both prepared cores from JKV and blades from Elsloo.

Keywords: Linear Bandkeramik, flint, procurement strategies, distribution networks, Graetheide, Rhineland.

INTRODUCTION

The Bandkeramik settlement Geleen-Janskamperveld (JKV), situated on the well-known Graetheide Plateau in the Netherlands (Fig. 1), was excavated in 1990/1991. An initial analysis of house plans, decorated ceramics and settlement structure is given by Louwe Kooijmans *et al.* (2003), and De Grooth (2003b) reported on the flint assemblage. The full excavation report will be published shortly (Van de Velde *et al.* 2007).

The settlement extended over approximately 4.5 ha, of which 2.7 ha, or 61% was investigated. Habitation started early in the Flomborn phase of the Linearbandkeramik (LBK). This first occupation lasted approximately 100 years, comprising the phases Ib and Ic of the Dutch chronology (cf. Modderman 1970). During this time, an estimated 90 houses were constructed, of which 69 have been excavated. Originally, on the basis of the decorated ceramics, Van de Velde distinguished five phases within this Early LBK habitation (Louwe Kooijmans *et al.* 2003). A subsequent analysis, combining data on decorated ceramics and on the structure of houses, arrived at a subdivision into four housing generations (Van de Velde *et al.* 2007). After a marked hiatus, the site was resettled during Modderman's phase IIc, but the extent and spatial organisation of this phase of occupation cannot be assessed.

The main objective of the study presented here was to investigate the way(s) flint working was organized at the Early JKV site, thus getting an insight into the procurement strategies and the technological choices made at the beginning of Bandkeramik settlement west of the river Rhine. For this purpose, a total of 7941 flint artefacts, weighing *c.* 58 kg, were analysed (De Grooth 2007). 4864 of these were recovered from 42 pits containing at least 15 flints, and dated to the Flomborn phase by their ceramic content (Table 1). These figures differ from those published previously (De Grooth 2003b), partly because material from some of the excavation trenches were not

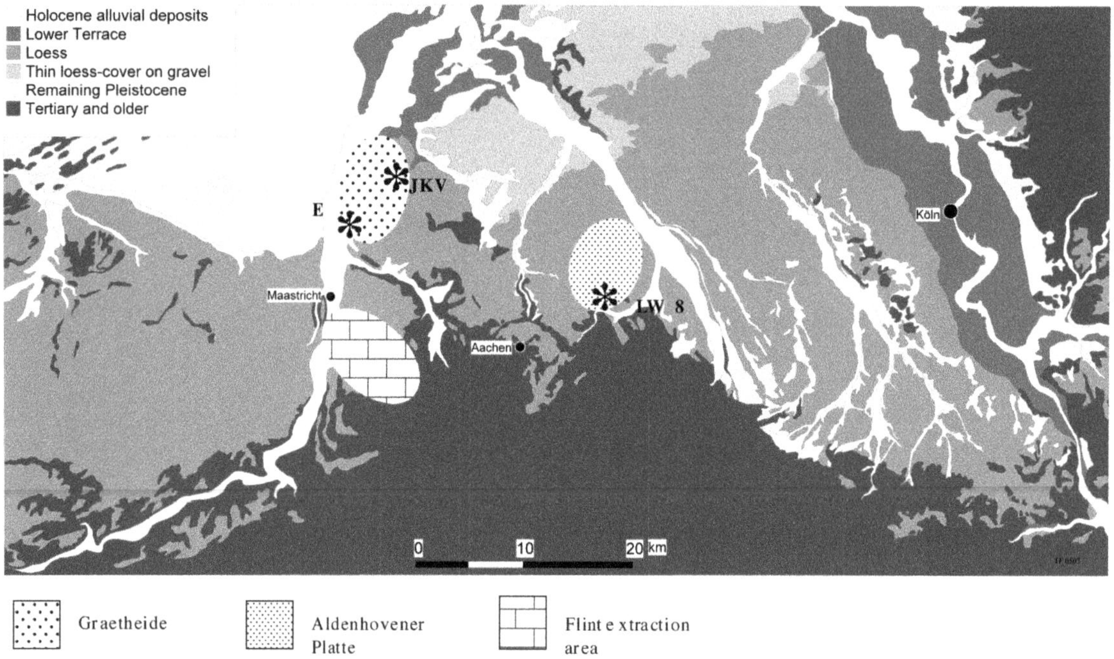

Fig. 1: The location of Geleen-Janskamperveld (JKV) and other sites under discussion. Map based on Modderman 1970, Fig. 1.

Type	JKVE1		JKVE2		JKVE3		JKVE4		All Early	
	N	%	N	%	N	%	N	%	N	%
Cores/hammerst	13	1.3	11	0.8	18	1.0	5	0.7	47	1.0
Hammerstone fragm	27	2.8	28	2.1	43	2.4	11	1.6	109	2.2
Crested blades	10	1.0	7	0.5	11	0.6	4	0.6	32	0.7
Rejuvenation flakes	22	2.3	29	2.1	46	2.5	9	1.3	106	2.2
Flakes with cortex	213	21.8	261	19.2	356	19.6	138	19.6	968	19.9
Flakes without cortex	354	36.2	466	34.2	693	38.1	219	31.1	1732	35.6
Blades with cortex	34	3.5	41	3.0	46	2.5	19	2.7	140	2.9
Blades without cortex	114	11.7	125	9.2	166	9.1	91	12.9	496	10.2
Tools	102	10.4	129	9.5	182	10.0	96	13.6	509	10.5
Chips (<15 mm)	55	5.6	244	17.9	226	12.4	98	13.9	623	12.8
Artificial blocks	33	3.4	22	1.6	32	1.8	15	2.1	102	2.1
Total	977	100.0	1363	100.0	1819	100.0	705	100.0	4864	100.0

Table. 1: Geleen-Janskamperveld: amounts and percentages of artefacts through time, for Early LBK habitation (ceramically dated pits containing at least 15 flint artefacts).

initially available for study and partly because some of the pits were assigned to a different phase after re-analysis of the decorated ceramics.

1. RAW MATERIAL PROCUREMENT

The flints used at JKV originated almost exclusively in the western facies of the Lanaye Member of the upper Cretaceous Gulpen Formation (Felder and Felder 1998; Felder *et al.* 1998, Felder and Bosch 2000), occurring in the limestone area south of the river Geul, at a distance of 20-30 km from the settlement. This region is considered to be the major procurement area for Bandkeramik settlements in the Graetheide cluster (Bakels 1978, de Grooth 1987), as well as for the Rhineland (e.g. Zimmermann 1995, Zimmermann *et al.* 2004). The Lanaye stratum comprises 23 different flint-bearing levels (numbered from bottom to top). Of these, layer 10, comprising the largest amount of usable flint, is the one most extensively exploited during the Neolithic. Lanaye flints (commonly referred to by archaeologists as Rijckholt flint), display a large variability. The nodules in general have a length, width and thickness of at least 30 cm. The cortex is thin, rough and whitish. Natural fracture planes are often covered with iron incrustations. The internal colour varies from very dark to very light grey, both sometimes with a hint of blue. The surface of the artificial fractures is smooth, but dull, the texture is mainly fine-grained. Both colour and texture may vary within individual nodules; the lighter grey parts often contain zonated areas, with gradual transitions. Sometimes the zone directly under the cortex is the darkest, with a smoother, more vitreous texture. The main types of inclusions are (Felder *et al.* 1998, De Grooth 1998):

-Concentrations of light (white or light grey) round specks (<1mm); isolated small (1–3) and medium-sized (3–10 mm) round or ovoid spots, light grey or white, with the same texture as the matrix.

-Small (1–3 mm), medium-sized (3 –10 mm) and large (>10 mm) ringed spots, round or irregular in shape, with a smooth whitish outer ring and a smooth or rough darker centre.

-Large (>10 mm) spots, round or irregular, abrupt border, light grey or whitish, with a texture rougher than the matrix.

-Large, vaguely lighter grey flecks.

-Concentrations of black round specks (< 1mm); small and medium-sized (between 1 and 10 mm) black or very dark grey round, ovoid or sickle-shaped spots; dark wisps or tendrils.

-Small angular cavities.

For the most part, flints from the other levels within the Western Lanaye Member have the same characteristics, notably as regards colour, texture and the presence of both dark and light spots and specks. The detailed study of four of them showed the following divergences: dense concentrations of small white spots are present in layer 01; the flints from layer 2A often contain seemingly overlapping inclusions, and fossils that are not completely silicified; abrupt transitions in colour occur in the laminated areas in layer 12A. Then, the cortex may be thick (e.g. layer 02A, 12A) or irregular, with hollows and protuberances (layer 01).
The combined presence of several kinds of dark inclusions is the major characteristic distinguishing Lanaye flints from other flint types in the region (De Grooth in prep.).

Lanaye flint nodules could have been extracted from four different depositional contexts:

-Firstly, from their primary position, in the chalk bedrock.

-Secondly, from slope, talus or scree deposits, that came into being when valleys developed during the Pleistocene cut into the chalk beds, thus exposing and eroding them.

-Thirdly, from residual loams (also known as eluvial deposits) that are the result of disintegration of the chalks during the Tertiary.

-Finally, from gravels deposited by the river Meuse during the Pleistocene and the Holocene.

The conditions prevailing in these secondary deposits sometimes led to visible alterations in the aspect of the flints. Flints from slope deposits differ from the material in primary context only in regard to a slight weathering of the cortex and other natural surfaces. Material collected in river gravels may be identified as such only when parts of the cortex or other natural surfaces are still present, as the river transport leads to heavy abrasion of cortex, to a decrease in size of the flints, and to an increase in the frequency of non-cortical natural surfaces. These often have a battered aspect and carry a glossy patina (pebble patina, Verhart 2000). The most extensive alterations, however, are present on flints embedded in residual loams, especially when these are mixed with iron-rich Oligocene sands.

They were found to differ from primary Lanaye material in the following aspects (De Warrimont and Groenendijk 1993; De Grooth in prep.):

-Differences in the colour and texture of the cortex.

-Changes in colour, as a result from the infiltration of iron compounds. This often manifests itself as a reddish brown zone directly under the cortex and/or as yellowish streaks that penetrate deeply into the nodules. In extreme cases, the whole nodule is affected.

-The presence of a thick (>1 mm) white zone directly under the cortex.

-Natural and artificial fracture planes with a 'dusty' aspect, due to the presence of dense concentrations of minuscule, vermiculate, spots.

-Concentrations of white specks (< 1mm) and small white spots (indicating the presence of flints from Lanaye layer 01 and/or the underlying Lixhe Member).

-The degree of translucency. Using a method devised by Ahler (1983), unweathered freshly knapped flakes were studied in a darkened room with a light source (provided by a 12 V/20W halogen desk lamp) diagonally behind them. The boundary between the opaque and translucent parts of the pieces was then marked in pencil, and the thickness measured with a pair of sliding callipers. Later on, the measurements were grouped into 5 classes: T1: translucency ≤ 2.4 mm (opaque); T2: translucency between 2.5 and 4.9 mm (low); T3: translucency between 5.0 and 7.4 mm (medium); T4: translucency between 7.5 and 9.9 mm (high); T5: translucency greater than 10.0 mm (very high). Whereas the material from the Rijckholt mine shafts and the slope deposits both displayed a low to medium translucency, those residual sites containing a mixture of loams and Oligocene sands produced highly translucent flints.

Prehistoric extraction points, that probably were in use during the Early Neolithic are known for three of the four depositional contexts (Fig. 2, Table 2).

-1. Lanaye nodules from both a primary context and from slope deposits may have been extracted in the mining area situated between Rijckholt (mun. Eijsden, NL) and Sint-Geertruid (mun. Margraten, NL). The well-known underground mining activities there are contemporary with the Michelsberg culture (Felder *et al.* 1998). Earlier mining of primary material, by means of shallow pits, cannot be excluded *a priori*. In the immediate vicinity, flints were extracted from slope deposits as well, as has been shown by the excavations in the Schone Grub, a dry valley in the north-western part of the prehistoric mining area (Rademakers 1998; Felder *et al.* 1998).

-2. The steep slopes between the Upper Terrace and the Middle Terrace surrounding the plateau known as De Kaap, located some 500 m further to the south, may also have been exploited (Felder 1998). At present these activities cannot be dated, but some Bandkeramik adzes have been found on the Rijckholt plateau (Brounen and Peeters 2000/2001).

Early Neolithic exploitation of residual flint deposits may have taken place at the following sites (cf. Felder 1998):

-3. Hoogbos, between Mheer (mun. Margraten, NL) and 's-Gravenvoeren (mun. Voeren, B). Exploitation is thought to have taken place in the steep valley slope, where the material crops out. Nothing is known about the age and character of the mining activities, but Louis (1936) reported that some of the cores and rejuvenation tablets collected here resemble those found at Omalien (i.e. Bandkeramik) sites in the vicinity of Liège. In colour, texture and inclusions this material is identical to the flints encountered at Rijckholt; the cortex, however, is rough, thin and brownish. Yellowish streaks are infrequently present, as are concentrations of small light spots. Opaque reddish brown zones are very rare. The translucency is low to medium.

-4. Banholtergrub, close to Banholt (mun. Margraten, NL), situated on the northern slope of a narrow dry valley. The eluvial deposits lie on the remaining chalks and are covered by Quaternary gravels. Some Oligocene sands are mixed with the loams (W.M. Felder, oral communication 29/06/2006). Recently, Brounen and Peeters (2000/2001) presented plausible evidence for (open-cast) mining and knapping activities at this site during Early Neolithic, more specifically Bandkeramik times. Again, the material is similar to the Rijckholt sample. The cortex, however, is rough, thin, brown or grey. Frequently a thin reddish brown, glass-like zone below the cortex is present, and a thick white layer may occur; brown or yellowish streaks are common, as are concentrations of light specks (<1 mm); the fracture surface lacks the dusty appearance typical for Rullen (see below). The translucency is medium to high.

-5. A cluster of extraction sites at Rullen and Sint Pietersvoeren (Vrouwenbos according to Felder 1998; also known as Bois Communal and Bois des Sapins) in the municipality of Voeren (B). The material extracted

	Primary context	Slope deposits	Residual loams	Oligocene sands	Oligocene cobbles
Rijckholt mines	X				
Rijck. slopes/Kaap		X			
Hoogbos			X		
Banholt			X	X	
Rullen c.s.			X	X	X
Rodebos			X	X	X

Table. 2: Geological characteristics of known extraction points.

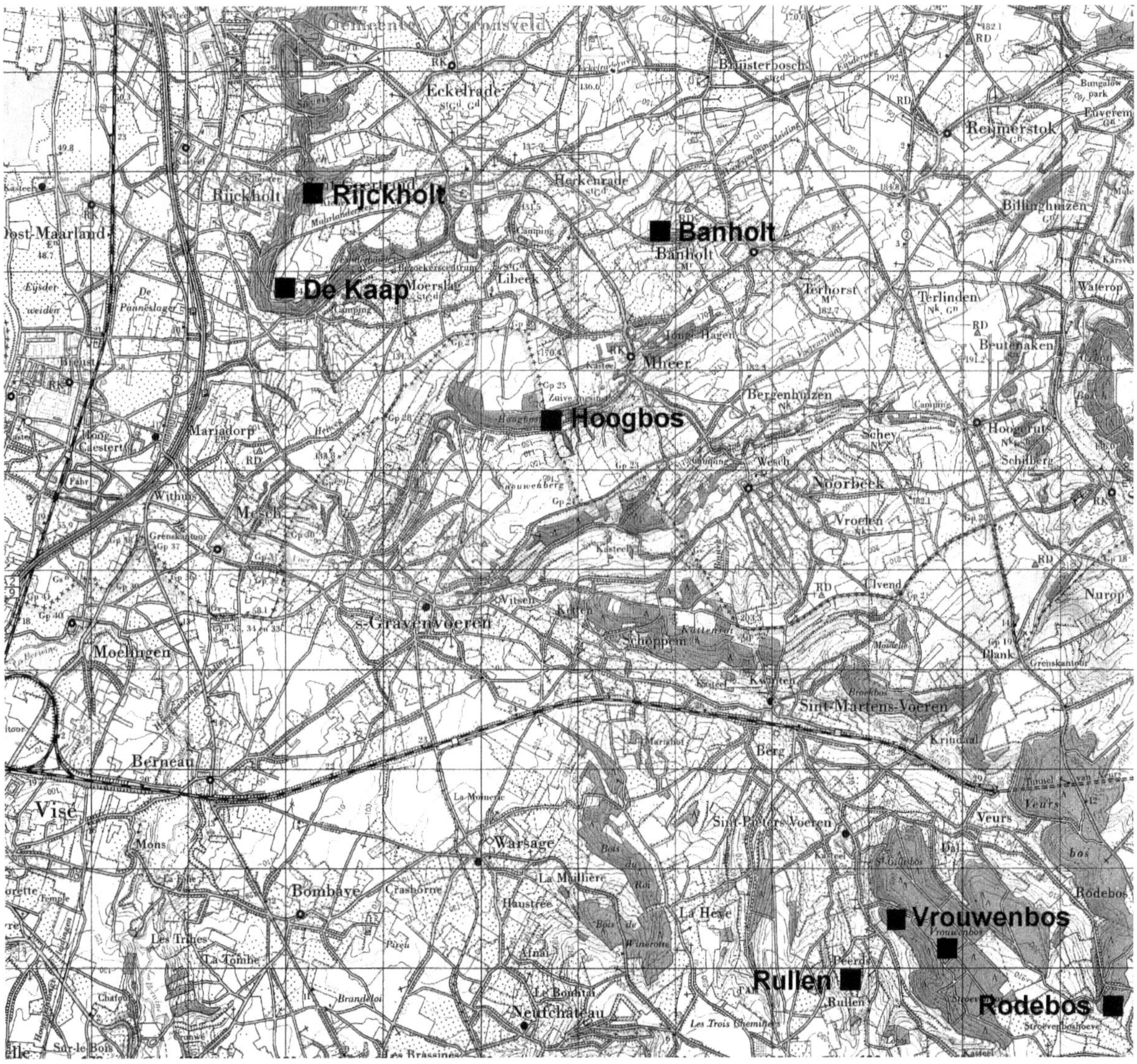

Fig. 2: Possible Early Neolithic extraction sites. Map: Zuid-Limburg, 1:50.000 (topografische dienst Nederland 1980).

here is commonly known as Rullen flint. At these sites, the residual loams are mixed with important amounts of Oligocene sands, rich in iron oxides, displaying intense red and yellow colours. In 1998, during a rescue excavation prior to the construction of a liquid gas pipeline, traces of a funnel-shaped extraction pit, with a preserved depth of ca. 300 cm and a reconstructed diameter of 7,20 m, were found at Rullen (Vermeersch *et al.* 2005). The available radiocarbon dates correspond to the end of the Neolithic or even later. The presence of blanks and tools in earlier Neolithic settlements, however, points to extensive extraction activities in this period: Rullen flints are encountered sporadically during the LBK, but are an important to predominant raw material in settlements of the Grossgartach, Planig-Friedberg and Rössen cultures (Gehlen and Schön in press). At the Rullen/Sint Pietersvoeren sites, materials originally deriving from all the levels within the Lanaye Member were exploited. This follows from the observation that concentrations of light specks (<1 mm), the size of a pin's head, and concentrations of small light spots are common, whilst abrupt transitions in colour and sharply defined stripes parallel to the cortex occur as well. The cortex is brown, sometimes rough, sometimes smooth; mostly thin, but sometimes thick. A thick, white layer is often present, especially in material from the Sint Pietersvoeren (Bois Communal and Bois des Sapins) sites. The fracture planes are dull and mostly have a 'dusty' aspect, probably because the bleaching has made the presence of dense concentrations of minuscule, vermiculate, spots visible. As befits their origin in the Western Lanaye deposits, the basic colour of the flints found at the different Rullen/ Sint Pietersvoeren extraction sites was grey. Three phenomena, however, caused alterations, that make it possible to distinguish these flints from the Lanaye material from primary chalk deposits. Firstly, their long stay in an eluvial matrix gave them a 'bleached' aspect, possibly caused by the dissolution of the carbonates that are present in large amounts in primary Lanaye material (cf. MacDonnell *et al.* 1997; Giot *et al.* 1986 describe the same loss of carbonates for flints from Le Grand-Pressigny). This bleaching sometimes affected the dark inclusions as well. Then, possibly because of the same loss of carbonate content, Rullen flints have a high translucency (predominantly class 4-5). Thirdly, infiltration of the iron compounds present in the matrix led to a yellowish-brown discoloration. At its most intense, this results in the 'honey' or 'egg-yolk' coloured nodules often seen as the typical Rullen flint (esp. Löhr *et al.* 1977). In many artefacts found at the Rullen and Sint Pietersvoeren sites, however, solid reddish- or yellowish-brown colours are limited to the outer part of the nodules, whilst the rest of the piece is grey, with just some brown, yellow or orange streaks.

-6. Rodebos, close to Remersdaal (mun. Voeren, B). The site is undated and the character of the mining activities unknown. The geological situation is identical to that at Rullen; i.e. residual loams mixed with Oligocene sands. Besides the basic Lanaye attributes, this material is characterized by the frequent presence of a reddish brown glassy zone under the cortex. Sometimes a thick white zone occurs as well. Specks, yellowish wisps and fracture planes with a dusty aspect are rare. The fracture surfaces are more reflective than those at Rijckholt and Rullen, and the translucency is extremely high: almost 60% of the sample belongs to class 4 and 5.

No specific extraction points for gravel flints are known, but the material was widely used during the Neolithic, especially in areas where it formed the nearest source of raw material (Weiner 1997; Mischka 2004).

By comparing the raw material characteristics of the different extraction points with those of the JKV flints, the latter's probable origin could be established.

Firstly, their slightly weathered cortex shows that the Lanaye flints worked by the inhabitants of Geleen did not originate from a primary geological position. Nor were they collected in river gravels. The majority seem to derive from a secondary position, either from slope deposits, such as found in the vicinity of the Rijckholt mines, or from residual loams.

Secondly, in a sample of 358 well-dated JKV artefacts, 82 pieces (23 %) showed a vitreous reddish-brown zone underneath the cortex; 50 times (14 %) a thick white zone was present under the cortex, and 16 artefacts (4.5%) had yellowish-brown streaks. The characteristic features of the Rullen flints (solid yellow or brown colours, concentrations of small light round spots and dusty fracture surfaces) were absent in the sample. Thus, the JKV material shows characteristics not encountered in the Rijckholt or Rullen samples, but resembles best the raw material collected at Banholtergrub and Rodebos (Table 3).

	White zone	Red zone	Brown streaks	Dusty surface
JKV	14 %	23 %	4.5 %	No
Rijckholt	No	No	No	No
Hoogbos	No	Rare	Rare	No
Banholt	Common	Common	Common	No
Rullen c.s.	Common	Frequent	Frequent	Frequent
Rodebos	Common	Frequent	Rare	Rare

Table. 3: Comparison of raw material characteristics between JKV and extraction points.

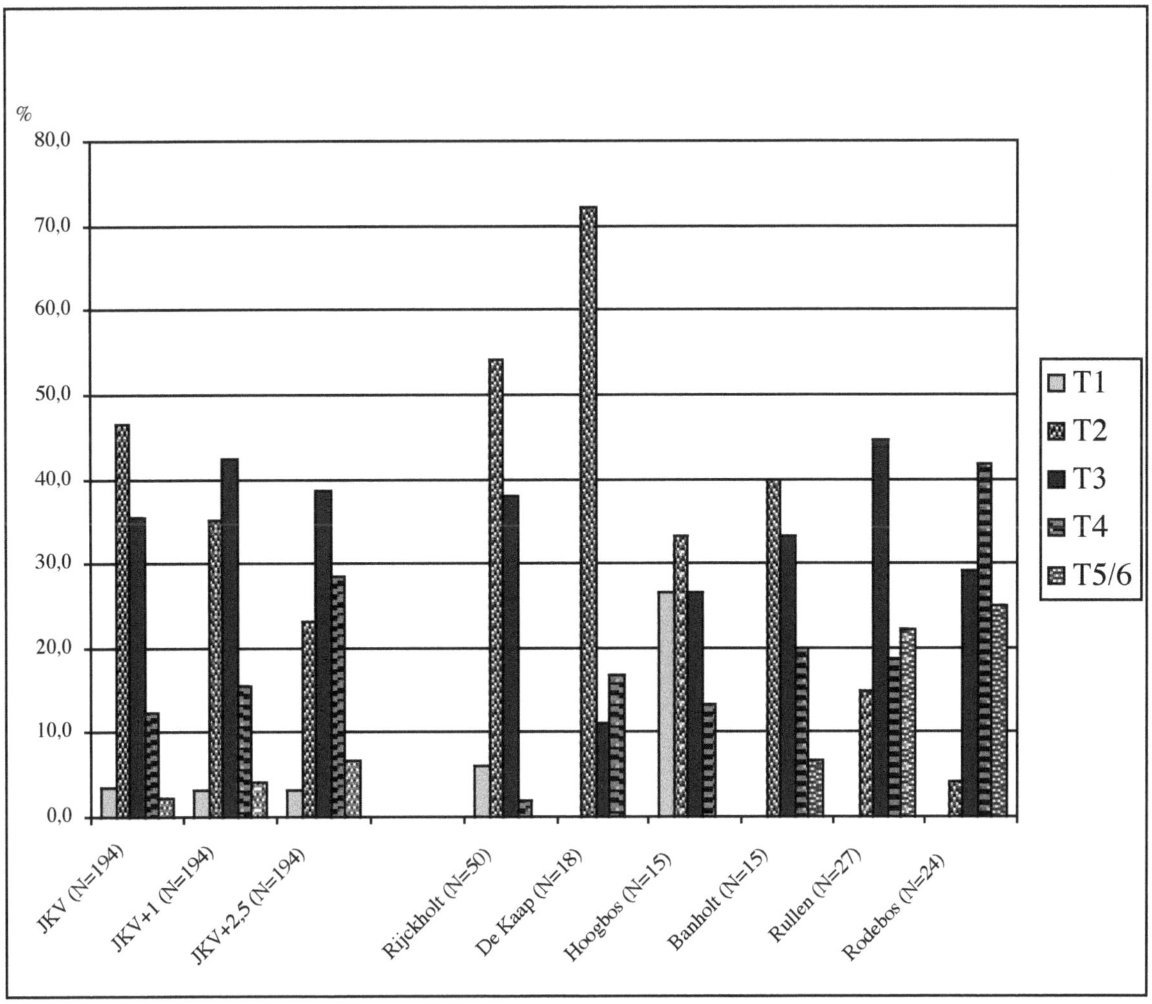

Fig. 3: Comparison of translucency between JKV and extraction points.

Finally, the translucency of 271 JKV artefacts was measured. 3% were opaque, 46% slightly translucent, 35 % showed a medium translucency and 16% were highly translucent. Moreover, 181 of the artefacts (67%) were completely translucent (Fig. 3). This phenomenon partly reflects the relatively low thickness of the JKV artefacts, but it also points to an overall high translucency of the raw material used. In order to make a comparison with the data from the extraction sites possible, the measured translucency was presented in three ways: 1. raw data; 2. increase with one mm for all entirely translucent artefacts; 3. increase with 2.5 mm. This, of course, leads to an increase of pieces in the higher translucency classes. Again, the JKV sample does not resemble the Rijckholt material, even if only the 'raw' data are considered. The correspondence with the data from Rullen, Banholt and Hoogbos is much better, and improves when regarding the enhanced data; Rodebos, on the other hand, seems a less likely source, because flints with a low or medium translucency are scarce at that location.

Thus, it seems highly probable that the majority of JKV raw material was collected at the Banholtergrub (or at one or more as yet unknown extraction sites, where the flints were embedded in similar conditions), with Hoogbos and Rodebos as possibilities of secondary importance.

The flints of the Bandkeramik sites at Elsloo and Beek-Kerkeveld display the same characteristics (observation by the present author). Given the mention of transparent reddish-brown zones as typical for the so-called 'Rijckholt' flint encountered elsewhere in the Bandkeramik world (e.g. Deutmann 1997, Löhr *et al.*1977, Zimmermann 1988, 606), it seems plausible that this material mainly originated from Banholt as well. It even reached as far east as Friedberg-Bruchenbrücken in Hesse, where at least two artefacts made of the Banholt variety were identified among the Oldest Bandkeramik material by A. L. Fischer and the present author, using the author's reference collection (Fischer 2005, De Grooth in prep.).

2. THE PROCESS OF FLINT WORKING AT GELEEN JANSKAMPERVELD

The *chaîne opératoire* (Pelegrin *et al.* 1988) chosen by JKV's inhabitants to transform their raw material into manufactured products strongly resembles the one practised at other regional Bandkeramik sites, such as Beek-Kerkeveld (De Grooth 1987), Liège-Place St. Lambert (Cahen 1984), or Verlaine-Petit Paradis (Allard 2005). The toolkit used by JKV's inhabitants is extremely conventional, both in composition and in morphology (Table 4). Arrowheads, borers (or rather: inserts for drilling machines), truncated blades, blades with lateral retouches, end-scrapers and side-scrapers are the main standardized tool types. Following common practice, to these are added the blades with intensive gloss, interpreted as sickle inserts, even though they often are not modified by intentional retouch. Standardized tools were almost exclusively made from blades – with the exception of end- and side-scrapers. Arrowheads, too, at JKV were mainly made from blades, although their small size often makes it difficult to determine the type of blank. Most retouched tools show a direct steep retouch, with only arrowheads sometimes displaying bifacial or inverse flat retouch.

The cores and hammerstones discarded at JKV are quite small (Table 5).The blades are rather stocky; three quarters (76.5%) of them have a length/width ratio between 1:2 and 1:3 (and 48.1 % between 1:2 and 1:2.5).

3. DIACHRONIC OBSERVATIONS

In all habitation phases, the percentage of unmodified flakes at JKV is very high (between 67.4% and 77.3%), as is the percentage of unmodified flakes with cortex (between 34 and 39 %). Nevertheless, the assemblage displays important diachronic variations (Table 6). It is not easy to assess these figures looking only at percentages, firstly because of their interdependence, and secondly because of the enormous difference in frequency between unmodified flakes and cores. Therefore, the relative proportions of main artefact categories are presented as well (Table 7).

In the first habitation phase (E1) the proportion of flakes to cores is considerably lower than in the other phases, as is the proportion of flakes to blades plus tools. Two main interpretations are possible for this phenomenon. In the first, the phase E1 assemblages have too few flakes for every core, and the situation in the other phases is regarded as standard. If, alternatively, the phase E1 situation is regarded as standard, the other phases would have too many flakes, i.e. too few cores.

In the first case, during phase E1 flakes would have been underrepresented at JKV because the preparation of (some

Types	Early habitation		All material	
	N	%	N	%
Arrowheads	42	8.6	74	8.3
Borers	28	5.7	47	5.3
End-scrapers	144	29.5	264	29.7
Side-scrapers	13	2.7	23	2.6
Truncated blades	41	8.4	62	7.0
Retouched blades	80	16.4	179	20.1
Sickle blades	80	16.4	131	14.7
Retouched flakes	25	5.1	44	4.9
Splintered pieces	17	3.5	37	4.2
(micro) Burins	4	0.8	5	0.6
Notches	8	1.6	11	1.2
Denticulates	2	0.4	4	0.5
Microliths	3	0.6	3	0.3
Quartiers d'orange	0	0.0	1	0.1
Non-sickle gloss	1	0.2	5	0.6
	489	100.0	890	100.1

Table. 4: JKV, the main tool types.

			Surface mm^2
Tablets	Length	Width	
	61	47	3039
Blade cores	Platform L	Platform W	
	46	40	1955
Blades/blade tools	Width	Thickness	
	18	5	90

Table. 5: JKV, estimate of the amounts of blades produced (median dimensions in mm).

	E1	%	E2	%	E3	%	E4	%	All	%
Cores	13	1.3	11	0.8	18	1.0	5	0.7	47	1.0
Flakes	671	68.7	1028	75.4	1364	75.0	475	67.4	3538	72.7
Blades	158	16.2	173	12.7	223	12.3	114	16.2	668	13.7
Tools	102	10.4	129	9.5	182	10.0	96	13.6	509	10.5

Table. 6: JKV, diachronic variability of main artefact types through time.

	E1	E2	E3	E4
Flakes : Blades+Tools	2.6	3.4	3.4	2.3
Flakes : Cores	51.6	93.5	75.8	95.0
Blades+Tools : Cores	20.0	27.5	22.5	42.0
Blades : Tools	1.6	1.3	1.2	1.2
Flakes : Tools	6.6	8.0	7.5	4.9
N	977	1363	1819	705

Table. 7: JKV, proportions of main artefact categories through time.

of the) cores had taken place elsewhere, e.g. at the extraction sites, whilst during the other phases unprepared cores were brought into the settlement. As this practice would not influence the intensity of blade and tool production at the site, one would expect the proportions of blades and tools to cores to be similar for all phases. This clearly not being the case, the alternative interpretation seems to fit better. As it is known that flints of the Lanaye type circulated widely throughout the Bandkeramik world, the most plausible scenario would be based on the assumption that production and use of cores was a local affair during phase E1, but that in the other phases some of the cores prepared at the settlement were not discarded there, but exported. Moreover, the fluctuations in the ratio of cores to blades plus tools, and of flakes to blades plus tools indicate that this export may have occurred at different stages in the reduction process as well. It is not easy to interpret these figures in more detail, partly because they may result from several different strategies, partly because of the many parameters involved. A tentative interpretation would be:

E1: local production; no cores, but some (un)modified blades were exported (hence the relatively low proportion of blades and tools in relation to both flakes and cores).

E2: local production, but some of the cores were exported (hence the higher ratio of flakes to cores), possibly some of them after preparation, some after initial blade production (resulting in a higher ratio of blades plus tools to cores).

E3: a part of the cores were exported after preparation (given the low ratio of blades plus tools to cores); the number of exported cores may have been somewhat lower than in the preceding phase.

E4: the number of exported cores increased again, but export mainly took place after initial blade production (hence the high ratio of blades plus tools to cores).

A corroboration of this interpretation is provided by a comparison of the size of blade cores and rejuvenation tablets discarded at JKV. The differences in surface of the striking platforms and of the complete rejuvenation tablets, allow for an estimate of the amount of blades to be made (Fig. 4). The median platform surface of cores is 1955 mm^2; the complete tablets have a median surface of 3039 mm^2. Thus, a median surface of 1084 mm^2 was removed during one stage of blade production. Given a median surface (i.e. width x thickness) of 90 mm^2 per blade, about 12 blades would have resulted (Table 5). Presuming that the same number of blades was produced before core rejuvenation, every core might have yielded some 25 blades. If most cores discarded at JKV originally produced blades, the proportions of cores to blades plus tools for the phases E2 and E3 would correspond quite nicely to this estimate, supporting the idea that some nodules were prepared as cores at JKV, but exported before blades were struck from them. In E1 the number of blades/tools is somewhat lower, and in E4 much higher than would be expected. Thus, during phase E1 some (un)modified blades may have been exported, whereas in E4 cores were used for blade production at JKV before leaving the settlement.

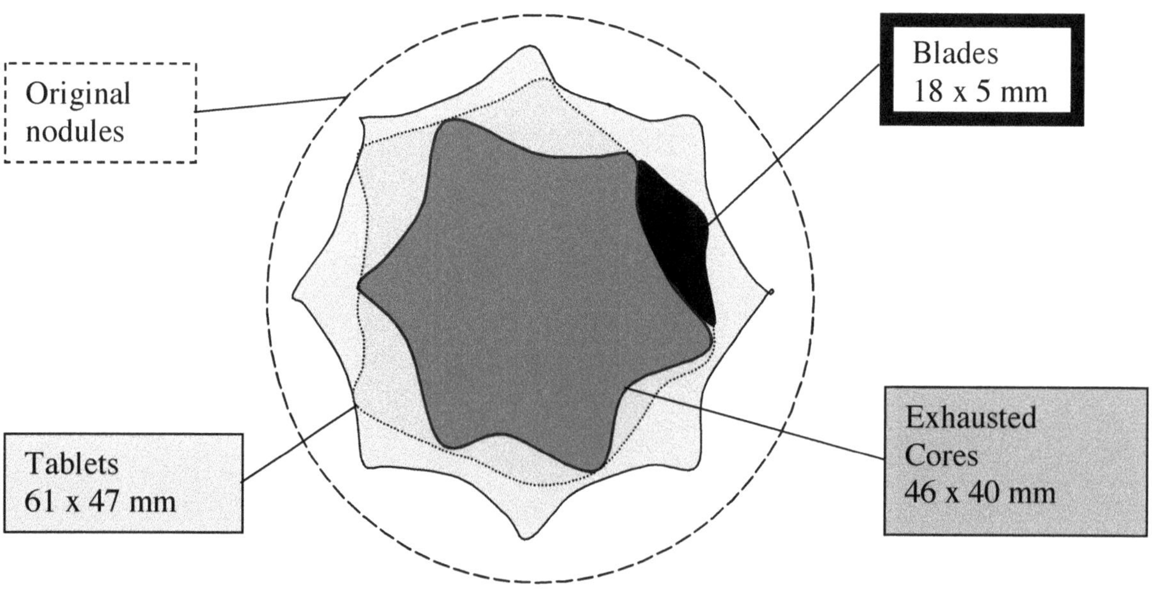

Fig. 4: JKV, relationships between dimensions of blade cores, rejuvenation tablets and blades.

4. INTER-SITE COMPARISONS

To assess the value of this interpretation, one would like to compare the JKV-data with those from the settlements that could have received its flints, especially sites further to the east that used important quantities of Western Lanaye flints, as well as from contemporary settlements in the Graetheide region, located at approximately the same distance from the extraction sites.

Unfortunately, at the moment is it impossible to synchronize the JKV chronology with the four micro habitation phases that comprise the Early LBK period at Elsloo (cf. Van de Velde 1979). On the other hand, I am fairly confident that, despite the differences in the analytic techniques applied, one may equate the JKV habitation phases as defined by Van de Velde (2007) with the House Generations (HG) of the Rhineland, as they are based on the analysis of quite similar decorated pottery. Both phase JKV-E1 and House Generation I belong to the beginning of Modderman's (1970) phase Ib, JKV's generation E4 is situated somewhere during Modderman Ic, as are House Generations IV and V, with HG VI and VII representing Modderman Id, when JKV had been abandoned (cf. D. Mischka 2004, Abb. 9).

The overall number of flint artefacts at LW 8 is considerable: 9532 (Zimmermann 1988, Abb. 576), of which 7554 are assignable to dated features (D. Mischka 2004, Abb. 15). The amount of flint recovered at Langweiler 8 and the other settlements on the Aldenhovener Platte during the Early LBK (comprising the first seven House Generations), however, is so small as to preclude a meaningful comparison on the level of House Generations (Table 8). At best one may observe that the percentage of unmodified flakes at JKV is consistently higher than that at Langweiler 8. Given the high percentage of unmodified flakes and of pieces with cortex, as well as a smaller amount of unretouched blades and tools than is found at neighbouring sites, Langweiler 8 is often interpreted as having played a central role in the production and distribution of flint cores and blanks (Kegler-Graiewski and Zimmermann 2003, Zimmermann 1995, 2002). In this early period, however, the differences with the presumed consumer settlements of the Middle Merzbach valley are not really clear cut. Obviously, it took time for Langweiler 8 to establish itself as a regional flint redistribution centre. Weisweiler 17 is also described as a flint producing central place (Kegler-Graiewksi and Zimmermann 2003; Zimmermann 2004). As this site provided only 214 datable flint artefacts (of which 126 belong to the Early LBK), it is difficult to assess this claim.

Therefore, the comparison had to be performed at a more general level, concentrating on JKV, Langweiler 8 and Elsloo, and looking at the Flomborn period as a whole. Of course, the data may not be used directly to infer differences in the ways flints were worked, without taking into account differences in excavation and sampling methods and in post-depositional processes. The three sites share two handicaps: firstly, the topsoil was removed mechanically, causing the loss of a considerable amount of material (mainly tools and blades), as witnessed by the excavation at Sittard. There, the manually removed topsoil contained 55% of the tools, 56% of the cores and hammerstones, but only 37% of the waste flakes and blades (Modderman 1958/59, 113). Secondly: the pits' contents were not systematically sieved, resulting in the possible under-representation of smaller artefacts (Gronenborn 1997). Elsloo and Langweiler 8 were investigated during rescue excavations, often under extreme time stress (Zimmermann 1987, 636), whilst JKV, as a student training project, could be investigated in a (somewhat) more leisurely way. Moreo-

JKV	E1	E2	E3	E4			
N % flakes	977 68.7	1363 75.4	1819 75.0	705 67.4			
LW8	I	II	III	IV	V	VI	VII
N % flakes	49 59.1	230 61.7	209 67.5	553 62.0	774 64.2	188 64.9	262 64.5
LW2, LW9, LB7							
N % flakes	38 57.9	116 64.7	1214 51.8	453 52.8	115 69.6	110 60.9	1248 59.1

Table. 8: Comparison of the number of flint artefacts and of unmodified flakes at JKV, Langweiler 8, and other Early LBK settlements in the Middle Merzbach valley (Langweiler 2, Langweiler 9, Laurensberg 7). Merzbach valley data according to D. Mischka 2004, Abb. 15.

ver, the preservation at JKV is considered to be better than at the other sites (Louwe Kooijmans *et al.* 2003). Both at Elsloo (observation by the present author) and at Langweiler 8 (Zimmermann 1988, 635) only *c.* 3% of the unmodified flakes have a length under 15 mm, against 18% at JKV. The mean weight of unmodified flakes also differs: 5.8 g. (or 7.0 g when the chips are excluded) for JKV *vs.* 9.7 g. for Langweiler 8 (no data on weight are available for Elsloo). These differences, however, could partly have to do with the intensity of flint working *per se* as well: where more preparation takes place, more chips and smallish flakes are produced. To neutralize the possible bias caused by JKV being more carefully excavated, its data will be presented with and without chips (Table 9). From this data a number of observations may be derived:

-The percentage of unmodified flakes at both JKV and Elsloo is considerably higher than that at LW8.

-The percentage of cores is lowest at JKV, and highest at Elsloo, with LW8 in an intermediate position.

-The ratio of flakes to cores at JKV is very much higher than at both LW8 and Elsloo.

-At JKV there are 14 blades and 11 tools for every core; at LW8 7.7 blades and 6.6 tools, and at Elsloo only 3.9 blades and 2.4 tools.

-At JKV seven (or six) flakes are present for every tool; at LW8 four and at Elsloo 10.5.

-At JKV there are 5.4 (or 4.4) flakes for every blade, at LW8 3.6 and at Elsloo 6.4

These observations may be interpreted in the following way:

The three sites under consideration all have very high percentages of unmodified flakes, and thus would qualify as sites where flint was worked locally. In terms of the models developed by Zimmerman (1995) to study exchange mechanisms on the Aldenhovener Platte, the difference in the percentage of unmodified flakes between JKV and Elsloo on the one hand and Langweiler 8 on the other hand, would indicate that the earlier stages of core reduction were better represented at the former settlements. In other words: some of the cores arriving at Langweiler 8 had been prepared elsewhere.

As JKV and Elsloo both have a very high percentage of flakes, the low index of flakes to cores at Elsloo cannot be explained by assuming that core preparation was not performed locally. Instead, I think that the Elsloo cores remained in the settlement, whilst part of the JKV cores were exported. Instead, Elsloo was an exporter of blades; witness its low ratio of blades plus tools to cores. This is corroborated by the high index of flakes to blades plus tools. If Langweiler 8 was a receiver of JKV cores, one would expect the flake/core index at this site to have been lower than that of Elsloo. The relatively high index can be understood, however, when taking into account that Langweiler 8 in its turn was an exporter of further reduced (and rejuvenated) cores (cf. Graiewski and Zimmermann 2003). Moreover, if the blades missing at Elsloo were in part

	JKV	JKV (no chips)	LW8	Elsloo
N	4864	4241	1351	3515
% flakes	72.7	68.7	63.2	76.0
% cores	1.0	1.1	2.3	2.9
% blades plus tools	24.2	27.8	32.8	18.9
Flakes : Cores	75.3	62.0	27.5	26.4
Blades: Cores	14.2	14.2	7.7	3.9
Tools: Cores	10.8	10.8	6.6	2. 4
Flakes : Tools	7.0	5.7	4.2	10.5
Flakes: Blades	5.3	4.4	3.6	6.4

Table. 9: Comparison of the intensity of flint working at JKV, Langweiler 8 and Elsloo during the Flomborn period (Langweiler 8: Zimmermann 1988, Abb. 596. Elsloo: De Grooth 1987).

		JKV Early	LW 8 Early (I-VII)	Ald. Platte, general	Hambach 8
Tablets	N	31	8	163	48
	Mean L (mm)	60.8	54.3	56.2	52.0
	Mean W (mm)	47.7	38.5	41.9	37.8
Cores	N	44	18	628	28
	Mean L (mm)	53.7	51.8	56.8	50.6
	Mean W (mm)	44.9	41.2	43.7	34.9
	Mean Th	34.0	28.3	32.1	22.7

Table. 10: Comparison of dimensions of rejuvenation tablets and exhausted cores. LW 8 Early: data made available by A. Zimmermann, Cologne; Aldenhovener Platte and Hambach 8 according to Hohmeyer 1997, Taf. 54, 59, 71, 72.

transferred to LW8, they help account for the low ratio of flakes to blades and tools, and the high ratio of blades plus tools to cores.

-A nice corroboration of the idea that JKV exported part of its cores to Langweiler 8 is provided by the size of the core rejuvenation tablets (Table 10). Whereas the exhausted cores of JKV and Langweiler 8 have similar sizes, the tablets discarded at Langweiler 8 are considerably smaller, as though they have served to rejuvenate cores at a later stage of the reduction sequence. The data for the Aldenhovener Platte as a whole and for Hambach 8 (some 10 km to the north-east) confirm the trend, especially as they comprise material from both the Older and the Younger LBK (when cores tend to be larger).

The debris recovered at the Banholt extraction site makes the pattern even more complicated, as it yielded not only exhausted blade cores, but also quite a number of rejuvenation tablets (collected by the present author). Apart from large polyhedral blade cores such as those published by Brounen and Peeters (2009/2001) which recall the cores worked at Beek-Kerkeveld and other Younger LBK sites, the extraction activities are undated. However, smaller cores, closely resembling the JKV material, are present as well.

Therefore it seems plausible that during some of the time, some of the material was brought into some of the settlements under consideration in the shape of blades produced at the extraction site, as was the case for instance in the Gäuboden area of South-eastern Bavaria (De Grooth 2003a). Thus, although the three settlements were located at approximately the same distance from the extraction sites, they used different procurement strategies, and Langweiler 8 was in part dependent on cores and blades from the Graetheide settlements (Fig. 5).

At first sight, this variability in procurement and exchange strategies is surprising, as Bandkeramik long-distance exchange networks generally are thought to be based on longstanding, stable kinship ties, ties carefully maintained from one generation to the next (e.g. Krahn-Schigiol 2005, Lech 2003). A general overview of Flomborn-period population dynamics may provide an explanation.

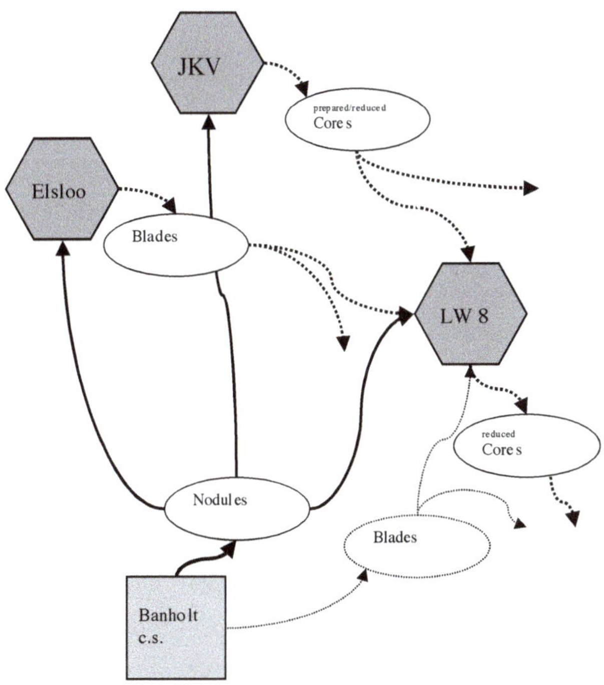

Fig. 5: Possible procurement and exchange strategies practised during the Early LBK in the Rhine-Meuse area. Thick continuous line: direct procurement of nodules. Thick dotted line: transport in the framework of down-the-line exchange. Thin dotted line: transport after direct procurement and production of blanks at extraction sites.

Radiocarbon and ceramic dates alike indicate that JKV was a 'first generation' settlement, as were Geleen-Kluis, Sittard, Elsloo and Stein (Van de Velde *et al.* in press). In other words, settlement west of the river Rhine was not a gradual, tentative step-by-step process, but started with a great leap westward, followed by filling-in of the areas in between during the next generations. Several sites in this Hinterland were settled in the same early stage as well, the best-studied being Langweiler 8 on the Aldenhovener Platte. Münch's (1999, 2005; cf. D. Mischka 2004, Zimmermann 2002) recent re-analysis of this area's decorated ceramics enabled her to give an exemplary insight in the dynamics of settling in the seven house generations of the Flomborn period for this settlement and its neighbourhood, comprising an area of some 25 square kilometres (Table. 11). Inhabitation began at Langweiler 8, with four contemporary

House generation	N settlements	N Houses
I	1	4
II	7	14
III	8	20
IV	9	23
V	8	24
VI	9	26
VII	9	29

Table. 11: Summary of settlement dynamics along the middle Merzbach-valley and its surroundings during the Flomborn period (after Münch 2005).

houses. In the next generation, not only did the number of houses there increase to 7, but at least six new settlements started, mostly inhabited by a single household. In the next house generations, some of these too grew in size, others remained single farmsteads, or were abandoned. In the last Flomborn generation, almost 30 houses are known to have been inhabited. Similar dynamics are documented for other parts of the Rhineland (Zimmermann *et al.* 2004, Claßen 2006), even Erkelenz-Kückhoven, located some 23 km to the north, on the northern fringe of the Rhenish loess area, was first settled in House Generation III (Lehmann 2004). At all of these sites, Lanaye flints, in part definitely of the Banholt variety, played an important role.

Given the fact that all settlements under consideration were located within one day's walking distance from the extraction sites, and could acquire alternative raw material of reasonable quality in local river gravels (Weiner 1997), there was no intrinsic need for the establishment of the exchange routines that were shown to exist between JKV (and other early Graetheide settlements) and the Rhenish settlements. On the contrary, the major incentive to maintain alliances with eastern neighbours and kin may have been the western settlers' need for a continuous supply of amphibolite and basalt adzes.

On the one hand, the pioneer situation outlined above would induce people to cherish and maintain kinship ties. On the other hand, this was a time of immense change in settlement and habitation, where one could not depend on traditional, fixed exchange networks alone, but had to be flexible and opportunistic. Yesteryear's trusted exchange partner and his family today may have moved on, or may be bound by obligations to other relations. In such an unstable situation, differentiated procurement and exchange strategies were called for, even on a micro-regional scale. Therefore, the fluctuations in the export of flint found at JKV, in my view, are perfectly compatible with the alternation of expansion and consolidation and the multiple networks visible in the Rhineland.

Marjorie E. Th. de Grooth
Aspelweg 49
D-53902 Bad Münstereifel-Eicherscheid
Deutschland
e-mail: grooth@t-online.de

REFERENCES

Ahler, S.A. 1983. Heat treatment of Knife River flint. *Lithic Technology* 12, 1-8.

Allard, P. 2005. *L'industrie lithique des populations rubanées du Nord-Est de la France et de la Belgique*. Internationale Archäologie 86. Rahden/Westphal: Verlag Marie Leidorf GmbH.

Bakels, C.C. 1978. Four Linearbandkeramik Settlements and their Environment. *Analecta Praehistorica Leidensia* 11.

Bohmers, A. and Bruijn, A. 1958. Statistische und graphische Methoden zur Untersuchung von Flintkomplexen IV. Das lithische Material aus den bandkeramischen Siedlungen in den Niederlanden. *Palaeohistoria* 6/7, 183-213.

Brounen, F.T.S. and Peeters, H. 2000/2001. Vroeg-neolithische vuursteenwinning en -bewerking in de Banholtergrubbe (Banholt, gem. Margraten). *Archeologie* 10, 133-150.

Cahen, D. 1984. Technologie du débitage laminaire. In M. Otte (ed.) *Les fouilles de la Place Saint-Lambert à Liège 1*, Liège: ERAUL 18, 171-199.

Cahen, D., Caspar, J.P. and Otte, M. 1986. *Les industries lithiques danubiennes de Belgique*. Liège, ERAUL 21.

Claßen, E. 2006. Die bandkeramische Siedlungsgruppe bei Königshoven. Unpublished PhD thesis, Cologne.

Deutmann, K.H. 1997. Feuersteinmaterial. In U. Boelicke, K.H. Deutmann, J. Lüning, F.W.V. Schmidt and P. Stehli, *Der bandkeramische Siedlungsplatz Aldenhoven 3, Kr. Düren*, in J. Lüning Hrsg., Studien zur neolithischen Besiedlung der Aldenhovener Platte und ihrer Umgebung, Rheinische Ausgrabungen 43. Köln/Bonn, 83-125.

Felder, P.J. and Felder, W.M. 1998. De geologie van de omgeving van het prehistorisch mijnveld te Ryckholt-St. Geertruid. In P.C.M. Rademakers (ed.), *De prehistorische Vuursteenmijnen van Ryckholt-St. Geertruid,* Maastricht, 111-137.

Felder, P.J., Rademakers, P.C.M. and de Grooth, M.E.Th. (eds.) 1998. *Excavations of Prehistoric Flint Mines at Rijckholt-St. Geertruid (Limburg, The Netherlands)*. Archäologische Berichte 12. Bonn.

Felder, W.M. 1998. Overzicht van de prehistorische vuursteenexploitaties binnen het Krijtgebied tussen Aken-Heerlen-Luik-Maastricht en Tongeren. In P.C.M. Rademakers (ed.), *De prehistorische Vuursteenmijnen van Ryckholt-St. Geertruid,* Maastricht, 169-193.

Felder, W.M. and Bosch, P.W. 2000. *Krijt van Zuid-Limburg. Geologie van Nederland deel 5.* Delft/Utrecht.

Fischer, A.-L. 2005. Die Silexartefakte der bandkeramischen Siedlung Bruchenbrücken, Stadt Friedberg/Hessen. Unpublished MA thesis. Cologne.

Gehlen, B. and Schön, W. (in press). Jüngere Bandkeramik – frühes Mittelneolithikum – Rössen im rheinischen Braunkohlenrevier: Steinartefakte als Spiegel einer sich verändernden Welt, *Rheinische Ausgrabungen.*

Giot, D., Mallet, N. and Millet, D. 1986. Les silex de la région du Grand-Pressigny (Indre-et-Loire). Recherche géologique et analyse pétrographique. *Revue Archéologique du Centre de la France* 25, 21-36.

Gronenborn, D. 1997. *Silexartefakte der ältestbandkeramischen Kultur.* Bonn: Universitätsforschungen zur Prähistorischen Archäologie 37.

Grooth, M.E.Th. de, 1987. The Organisation of Flint Tool Manufacture in the Dutch Bandkeramik. *Analecta Praehistorica Leidensia* 20, 27-52.

Grooth, M.E.Th. de 1998. Archeologische beschrijvingen van Ryckholt-vuursteen. In P.C.M. Rademakers (ed.), *De prehistorische Vuursteenmijnen van Ryckholt-St. Geertruid,* Maastricht, 160-163.

Grooth, M.E.Th. de 2003a. Dealing with Bandkeramik cherts. Procurement strategies in south-eastern Bavaria. In L. Burnez-Lanotte (ed.). *Production and Management of Lithic Materials in the European Linearbandkeramik. Acts of the XIVth UISPP Congress, University of Liège, Belgium 2-8 September 2001,* BAR International series 1200. Oxford, 37-44.

Grooth, M.E.Th. de 2003b. They do things differently there. Flint working at the Early Bandkeramik settlement of Geleen-Janskamperveld (The Netherlands). In J. Eckert, U. Eisenhauer and A. Zimmermann (eds.), *Archäologische Perpektiven; Analysen und Interpretationen im Wandel; Festschrift für Jens Lüning zum 65. Geburtstag,* Internationale Archäologie: Studia honoria Bd. 20. Rahden/Westf, 401-406.

Grooth, M.E.Th. de (in prep). Distinguishing Upper Cretaceous flint types exploited during the Neolithic in the region between Maastricht, Tongeren, Liège and Aachen. In J. Meurers-Balke and W. Schön (eds.) *Gedenkschrift für Jürgen Hoika.* Bonn: Archäologische Berichte.

Grooth, M.E.Th. de 2007. Flint working in the Bandkeramik settlement. In P. van de Velde *et al.* Excavations at Geleen-Janskamperveld, *Analecta Praehistorica Leidensia* 39.

Kegler-Graiewski, N. and Zimmermann, A. 2003. Exchange systems of stone artefacts in the European Neolithic. In L. Burnez-Lanotte (ed.), *Production and Management of Lithic Materials in the European Linearbandkeramik. Acts of the XIVth UISPP Congress, University of Liège, Belgium 2-8 September 2001*, BAR International Series 1200. Oxford, 31-35.

Krahn-Schigiol, Chr. 2005. Regionale und überregionale Kommunikationsnetzwerke der bandkeramischen Siedlungen auf der Aldenhovener Platte. In J. Lüning, Chr. Frirdich, A. Zimmermann (eds.), *Die Bandkeramik im 21. Jahrhundert. Symposium in der Abtei Brauweiler bei Köln vom 16.9-19.09.2002. (Internationale Archäologie: Arbeitsgemeinschaft, Symposium, Tagung, Kongress; Bd 7).* Rahden/Westf., Posterbeilage 1.

Lech, J. 2003. Mining and siliceous rock supply to the danubian early farming communities (LBK) in eastern central Europe: a second approach. In L. Burnez-Lanotte (ed.). *Production and Management of Lithic Materials in the European Linearbandkeramik. Acts of the XIVth UISPP Congress, University of Liège, Belgium 2-8 September 2001*, BAR International series 1200. Oxford, 19-30.

Lehmann, J. 2004. Die Keramik und Befunde des bandkeramischen Siedlungsplatzes Erkelenz-Kückhoven, Kreis Heinsberg (Grabungskampagnen 1989-1994). In H. Koschik (ed.), *Der bandkeramische Siedlungsplatz von Erkelenz-Kückhoven, Kreis Heinsberg I. Archäologie*, 1-364. Rheinische Ausgrabungen 54. Mainz.

Löhr, H., Zimmermann, A. and Hahn, J. 1977. Feuersteinartefakte. In R. Kuper, H. Löhr, J. Lüning and A. Zimmermann, *Der bandkeramische Siedlungsplatz Langweiler 9, Ge. Aldenhoven, Kr. Düren, Beiträge zur neolithischen Besiedlung der Aldenhovener Platte 2.* Rheinische Ausgrabungen 18. Bonn, 131-266.

Louis, M. 1936. Les ateliers néolithiques du 'Hoog Bosch' (Fouron-le-Comte, province de Liège), *XVIe Congrès International d'Anthropologie et d'Archéologie Préhistorique*, Bruxelles 1-8 septembre 1935, 348-357.

Louwe Kooijmans, L.P., van de Velde, P. and Kamermans, H. 2003. The early Bandkeramik settlement of Geleen-Janskamperveld: Its intrasite structure and dynamics. In J. Eckert, U. Eisenhauer and A. Zimmermann (eds.), *Archäologische Perpektiven; Analysen und Interpretationen im Wandel; Festschrift für Jens Lüning zum 65. Geburtstag,* Internationale Archäologie. Studia honoria Bd. 20, Rahden/Westf, 373-398.

McDonnell, R.D, Kars, H. and Jansen, J.B.H. 1997. Petrography and Geochemistry of Flint from Six Neolithic Sources in Southern Limburg (The Netherlands) and Northern Belgium. In: A. Ramos-Millán and M.A. Bustillo (eds.), *Siliceous Rocks and Culture,* Granada, 371-385.

Mischka, D. 2004. Zentraler Ort oder Nebensiedlung? – Die Feinchronologie der Grundformspektren des bandkeramischen Fundplatzes Kückhoven im Vergleich. In H. Koschik (ed.), *Der bandkeramische Siedlungsplatz von Erkelenz-Kückhoven, Kreis Heinsberg I. Archäologie,* Rheinische Ausgrabungen 54. Mainz, 537-594.

Modderman, P.J.R., 1958/1959. Die bandkeramische Siedlung von Sittard. *Palaeohistoria* 6-7, 33-120.

Modderman, P.J.R., 1970. Linearbandkeramik aus Elsloo und Stein. *Analecta Praehistorica Leidensia* 3.

Münch, U. 1999. Zur Siedlungsstruktur der Flombornzeit auf der Aldenhovener Platte. Unpublished MA thesis, Cologne.

Münch, U. 2005. Zur Siedlungsstruktur der Flombornzeit auf der Aldenhovener Platte. In J. Lüning, Chr. Frirdich, and A. Zimmermann (eds.), *Die Bandkeramik im 21. Jahrhundert. Symposium in der Abtei Brauweiler bei Köln vom 16.9-19.09.2002. (Internationale Archäologie: Arbeitsgemeinschaft, Symposium, Tagung, Kongress; Bd 7).* Rahden/Westf, Posterbeilage 2.

Pelegrin, J., Karlin, C. and Bodu, P. 1988. 'Chaînes opératoires': un outil pour le préhistorien. In J. Tixier (ed.), *Technologie préhistorique,* Notes et monographies techniques du CRA 24. Paris, 37-53.

Velde, P. van de, *et al.* (2007). Excavations at Geleen-Janskamperveld. *Analecta Praehistorica Leidensia 39.*

Verhart, L.B.M., 2000. *Times fade away. The neolithization of the southern Netherlands in an anthropological and geographical perspective.* Leiden: Archaeological Studies Leiden University 6.

Vermeersch, P.M., Chow, J., Creemers, G., Masson-Loodts, I., Groenendijk, A.J and DeBie, M. 2005. Neolithische vuursteenontginning op de site van Rullen (Voeren, prov. Limburg). In I. In 't Veld and W. De Clercq (eds), *Een lijn door het landschap. Archeologie en het vTn-project 1997-1998,* Archeologie in Vlaanderen 5. Brussel, 313-328.

Warrimont, J.P. de and Groenendijk, A.J. 1993. 100 jaar Rullen vuursteen: een kleurrijke vuursteensoort nader bekeken, *Archeologie in Limburg* 57, 37-46.

Weiner, J. 1997. Die Maasschotter der Niederrheinischen Bucht als Feuersteinlieferant für die bandkeramischen Siedlungsplätze Langweiler 8 und Langweiler 9. In J. Lüning, (ed.), *Studien zur neolithischen Besiedlung der Aldenhovener Platte und ihrer Umgebung (Beiträge zur neolithischen Besiedlung der Aldenhovener Platte 6),* Rheinische Ausgrabungen 43. Köln/Bonn, 599-646.

Zimmermann, A. 1988. Steinmaterial. In U. Boelicke, D. von Brandt, J. Lüning, P. Stehli and A. Zimmermann, *Der bandkeramische Siedlungsplatz Langweiler 8, Gemeinde Aldenhoven, Kreis Düren (Beiträge zur neolithischen Besiedlung der Aldenhovener Platte III),* Rheinische Ausgrabungen 28. Bonn, 569-787.

Zimmermann, A. 1995. *Austauschsysteme von Silexartefakten in der Bandkeramik Mitteleuropas.* Bonn: Universitätsforschungen zur prähistorischen Archäologie 26.

Zimmermann, A. 2002. Landschaftsarchäologie I: Die Bandkeramik auf der Aldenhovener Platte. *Bericht der Römisch-Germanischen Kommission* 83, 17-39.

Zimmermann, A. 2006. Production and demand of flint artefacts in the Bandkeramik of western Germany. In P. Allard, F. Bostyn and A. Zimmermann (eds.), *Contribution des matériaux lithiques dans la chronologie du Néolithique ancien et moyen en France et dans les régions limitrophes. Actes de la Xème Session de l'EAA, Lyon Septembre 2004,* BAR International series 1494. Oxford: 80-83.

Zimmermann, A., Richter, J., Frank, Th. and Wendt, K.P. 2004. Landschaftsarchäologie II – Überlegungen zu Prinzipien einer Landschaftsarchäologie. *Bericht der Römisch-Germanischen Kommision* 85, 39-96.

ACKNOWLEDGEMENTS

Many thanks to Thomas Frank and Andreas Zimmermann (Cologne) for Figure 1, and for providing unpublished data on the Langweiler 8 tablets; to my Leiden University colleagues Annelou van Gijn, Leendert Louwe Kooijmans, Piet van de Velde and Ivo van Wijk for their help in dealing with JKV's flints; to Rengert Elburg, Jean-Pierre de Warrimont and Kim Groenendijk, Joep Franzen, Henk and Eva Kars, Leo Verhart, Pierre Vermeersch, and Werner Felder and his fellow members of the Working Group on Prehistoric Flint Mining for stimulating discussions on flint characterization; and last, but certainly not least, to Paul Wheeler for improving the English text.

AN ECONOMY OF SURPLUS PRODUCTION IN THE EARLY NEOLITHIC OF HESBAYE (BELGIUM): BANDKERAMIK BLADE DEBITAGE AT VERLAINE 'PETIT PARADIS'

Pierre Allard, Laurence Burnez-Lanotte

Abstract: Research conducted on a lithic assemblage from the Early Neolithic at Hesbaye led us to explore the phenomenon of flint blade blank production with an original management strategy. This paper addresses a particular aspect of this complex economic and social phenomenon, which can be summarized by the following question: how can we qualify the production of blades in this precise case, and can we formalize and demonstrate the hypothesis of an 'overproduction', as we have already proposed. The problem of quantifying debitage products in concentrations, houses, and entire villages can be addressed at Verlaine based on the analyses that we propose here. The question is clearly not to quantify the exact number of blades produced, such an objective being inaccessible in our current state of research, but rather to formulate acceptable estimates. The method to evaluate the total number of blades requires several steps. In this paper, we shall present only the first attempts, based upon a few cases. The test we propose is based on the integration of three methods applied to the artefacts of a debitage concentration and detritic waste found within the same structure.

Keywords: Neolithic, flint, Bandkeramik, overproduction.

INTRODUCTION

Research conducted on a lithic assemblage from the Early Neolithic at Hesbaye led us to explore the phenomenon of a flint blade blank production with an original management strategy. This paper addresses a particular aspect of this complex economic and social phenomenon, which can be summarized by the following question: how can we qualify the production of blades in this precise case, and can we formalize and demonstrate the hypothesis of an 'overproduction' such as we have already proposed (Burnez-Lanotte and Allard 1998, 2003). What method can we use to estimate the quantity of blades produced? This implies estimating the number of blades produced by the best documented laminar productions of the settlement, and establishing a quantitative relationship between this calculation and the number of blades created by this production and used in the village. On this basis, we shall attempt to define the notion of an economy of lithic surplus production, which corresponds to an original type of blank production management.

1. THE ARCHAEOLOGICAL CONTEXT OF THE PRODUCTION

The site of Verlaine is located 15 km from Liège, Belgium (Fig. 1). From 1996 to 2002, L. Burnez-Lanotte, University of Namur, conducted an exhaustive excavation of the site, confirming the presence of a Bandkeramik (or Linear Pottery Culture) settlement extending over more than three hectares. The occupation includes several houses, 100 pits and nearly 22 flint debitage concentrations. In 2001, four debitage concentrations were discovered in the lateral pits of two joined houses. The concentration of

Fig. 1: Verlaine ' Petit Paradis', Belgium. Plan of the site 1996–2001; black stars indicate the debitage concentrations (Burnez-Lanotte, Allard 2003).

structure 130 was perfectly preserved in the bottom of the pit (Burnez-Lanotte and Allard op. cit.). A detailed technological study of this concentration, including exhaustive refitting, showed the particularities of the Verlaine artefact assemblage, whose debitage objectives seem to be related to an economic phenomenon of flint blade production in larger than usual quantities. Our analyses show that these products were exported and integrated within a circulation network covering the northern part of the Bandkeramik zone in Western Europe (Burnez-Lanotte and Allard op. cit., Allard 2005, 2007).

2. BLADE DEBITAGE AT VERLAINE: THE SIX CHARACTERISTIC ELEMENTS OF THIS TECHNICAL PRODUCTION

We have already presented elsewhere the particularities of blade production at Verlaine (Allard and Burnez-Lanotte 2007). They include a number of quantitative and qualitative factors that can be summarized as follows:

- The quantity of lithic artefacts at Verlaine is higher than at other sites currently excavated. In a preliminary study of 13 pits (including 6 lithic concentrations), we have already recorded nearly 60 000 objects including more than 1 000 blade cores. All the structures in the village contain laminar debitage waste products; the concentrations differ only in the form of their rejection and their exclusive composition of laminar debitage waste products (absence of tools, burned pieces, flake debitage, etc.). This spatial distribution indicates that there is no specific location for blade production activities and that this production is distributed among all of the households.
- The nearly exclusive procurement of raw materials from sources in direct proximity to the site.
- A laminar production with a single intention oriented towards blades of average to fine regularity, with parallel edges and a rectilinear profile. Blade length is standardized, as is clearly shown by the core dimensions. The intended range is 80–120 mm in length, 15–30 mm in width and 4–8 mm in thickness.
- A rational debitage management. The extraction of blades is organized according to two principal procedures. The most frequent consists of removing blades from the two edges of the debitage surface, in order to extract a central blade, resulting in a '2-1-2' rhythm. This rhythm was observed on over 50% of the cores (Allard 2005). The second procedure corresponds to a successive rhythm that begins on one side of the core and crosses the entire debitage surface toward the other side (rhythm 1-2-3 or 3-2-1). The first procedure, in association with the calibration of the products, is indicative of standardized and well controlled series (Binder and Gassin 1998).
- A high skill level. Though simple, this debitage of short blades by indirect percussion was performed by good knappers. The absence, or small amount of butt preparation, the large size of the core tablets, the complex preparations, as well as multiple opposed debitage sequences, are all signs of a high general knapping standard.
- Reduction sequences adapted to an intended production, aiming at maximal production. The refits of concentration 130 show that the management of volumes was oriented uniquely toward the production of standardized blades, 8 to 12 cm long. The debitage of blocks more than 20 cm long is particularly significant; this could have allowed for the production of longer blades, but we observe instead the successive debitage of multiple blade surfaces with changes in the volumes exploited and sequences of partial or total re-preparation of the block, to insure an optimal production of standardized blades (Allard 2007).

These different factors, compared to what is known at other regional and extra-regional Bandkeramik settlements, allow us to define a new production economy. In the context of studies that we are currently conducting to document this phenomenon, we address here the problem of quantifying the objects produced and consumed on the scale of the spatially distinct production units that correspond to concentrations, houses and villages.

3. METHOD TO QUANTIFY LAMINAR PRODUCTION: PRELIMINARY CONSIDERATIONS

Any attempt to quantify the products yielded by a blade core is a particularly perilous task. We must first emphasize the major difficulties associated with this calculation:

- Flint nodules, Campanian flint in this case, have highly variable forms and dimensions; the volumes of the blocks to be knapped thus cannot be considered as equivalent. Therefore, calculations of the number of blades produced by each block must be made case by case, based on exhaustive refits.
- The quality and homogeneity of the raw materials is

highly significant, since imperfections or irregularities in the flint directly influence the core preparation and reduction, and consequently the quantity and quality of the blades produced. This problem increases if several distinct flint types are exploited.

- The nature of the production is crucial: indeed, if the objectives do not tend toward standardized products, at least in terms of dimensions, their quantification becomes too speculative.

- Finally, the uniform application of a debitage technique is also essential. According to the results of experiments conducted by Jacques Pelegrin (personal communication), pressure debitage produces the most standardized products, though indirect percussion also produces fairly standardized blanks.

4. THE QUANTIFICATION OF DEBITAGE PRODUCTS AT THE SCALE OF SPATIAL UNITS

The problem of quantifying debitage products in concentrations, houses, and entire villages can be addressed at Verlaine based on the analyses that we propose here. The problem is clearly not to quantify the exact number of blades produced, such an objective being inaccessible in our current state of research, but rather to formulate acceptable estimates.

4.1. Characterization of waste products in 'concentrations'

The data from concentrations at Verlaine present characteristics that justify their quantitative exploitation:

a) These concentration waste products constitute 'closed productive assemblages', which have the following characteristics:

1- Homogeneity of rejected objects and their burial (at least for the smallest pieces): the concentration of lithic remains is homogeneous, regardless of the archaeological structure in which they are contained (only flint, no mix with other materials);

2- Spatial and temporal homogeneity of the production: the abandoned concentration corresponds to a strictly coherent production, both temporally and spatially;

3- High standardization of the blanks produced: the blades are produced in series, and most correspond to recurrent standards (see d) below).

b) Among the well preserved concentrations, extensive refits could be made in structure 130. The proportion of refit pieces over 2 cm is 29% (1243 out of 4272), representing 75% of the weight of the assemblage (74,600 kg for 98,513 kg total weight).

c) Only one type of flint was included in the concentrations: the fine-grained, gray Campanian flint of Hesbaye. However, this flint is not always of good quality: its numerous imperfections sometimes presented difficulties, or even interrupted the core reduction process.

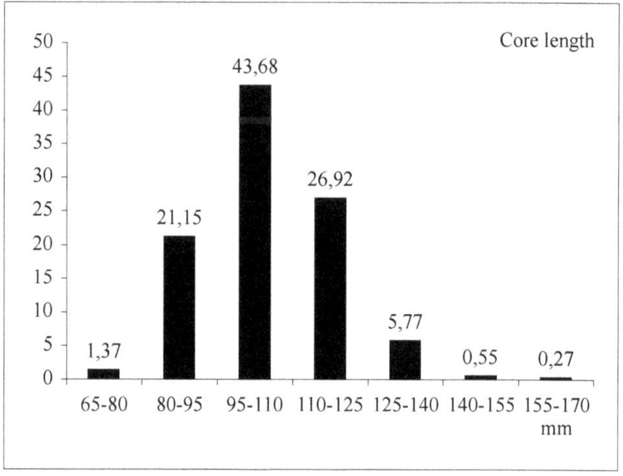

Fig. 2: Verlaine 'Petit Paradis'. Dimensions of the cores of feature 1.

d) The dimensions of the blade products, and especially the cores, are highly standardized. Among the 439 cores in structure 1, 85% were abandoned with a blade surface of 8 to 12 cm and 44% at 9.5 to 11 cm (Fig. 2).

e) The debitage of blade products was made by indirect percussion (Allard *op.cit.*). The macroscopic stigmata are particularly clear in this assemblage. The principal characteristics we observe are the absence, or only slight preparation, of the striking platform, often leaving a protruding edge, smooth and sometimes concave butts, a striking angle of 80-90° and the frequent appearance of a half-circle behind the butt.

f) The refits and debitage characteristics show that these were good knappers, notably in their ability and flexibility to adapt the optimal debitage of standardized blades to the initial morphology of the flint nodules, the low number of rejuvenation flakes per block, and especially in their responses to imperfections in the flint (Allard 2005, 2007). We thus observed efficient and regularized block reductions performed by experienced knappers.

4.2. The methods of production calculation applied to the concentrations

Our reflections were based on the refits of the concentration of structure 130. There are few complete refits, but the partial ones, notably those associated with repairs to core striking platforms, are systematic, except of course for the blocks with natural striking platforms.

The method to assess the total number of blades follows several steps. In this paper, we present only the first attempts, which are based on a few cases. In optimal situations, we applied three distinctive approaches, but more generally two. At this stage of analysis, only one has been systematically developed. These calculations are based on the laminar debitage volume observed through refits of core tablets, and the opening of striking platforms on cores (which shows in negative the maximum volume exploited). From a methodological perspective, the refits that correspond to repairs and the opening of striking platforms are sufficiently pertinent to enable us to attempt an estimation of the number of blades produced. In addition, they are advantageous because they are relatively easy to find in all the concentrations, even the most eroded. However, the negative volume visible between the opening of the striking platform and the core or core tablet corresponds to the total volume exploited, including the small preparation flakes, platelets, and other products. It thus corresponds not only to the standardized blades, but to all laminar debitage products.

Experimental debitage (method 1)

The experimental debitage was carried out by Jacques Pelegrin (CNRS-UMR 7055). He worked with a block of fine-grained Hesbaye flint, around 15 cm long, collected during the excavation of test pits in the dry valley that runs through the Bandkeramik site of Verlaine. The laminar volume corresponds to that estimated for the archaeological pieces (products 8-12 cm long). The debitage began with the creation of a frontal crest, and developed over a surface of an average of five blades. A core tablet was removed after a first production phase of around 30 laminar products. The five last blades were short and flat. The products were measured, and compared with the blades in concentrations and other structures, as well as with the tools from Verlaine (Allard 2005). This debitage served as first unit of measure to estimate the number of products obtained from the cores of concentration 130.

This estimation is complicated by two problems:
The experimental blades are slightly narrower and thinner than the retouched products of the Verlaine structures (Fig. 3). In fact, their dimensions correspond to those of the laminar products found in the concentrations, i.e. to the totality of rejected blades and broken standardized blades. It is thus probable that the number of experimental blades is greater than that of the Bandkeramik cores.
The quality of the experimental debitage shows that the standardized blades were removed one after the other with few maintenance products or small correction bladelets. This is due to the excellent initial preparation of the block. At Verlaine, the preparations are generally less thorough, and the maintenance products are more numerous. This probably means that the number of experimental blades is a little too high.

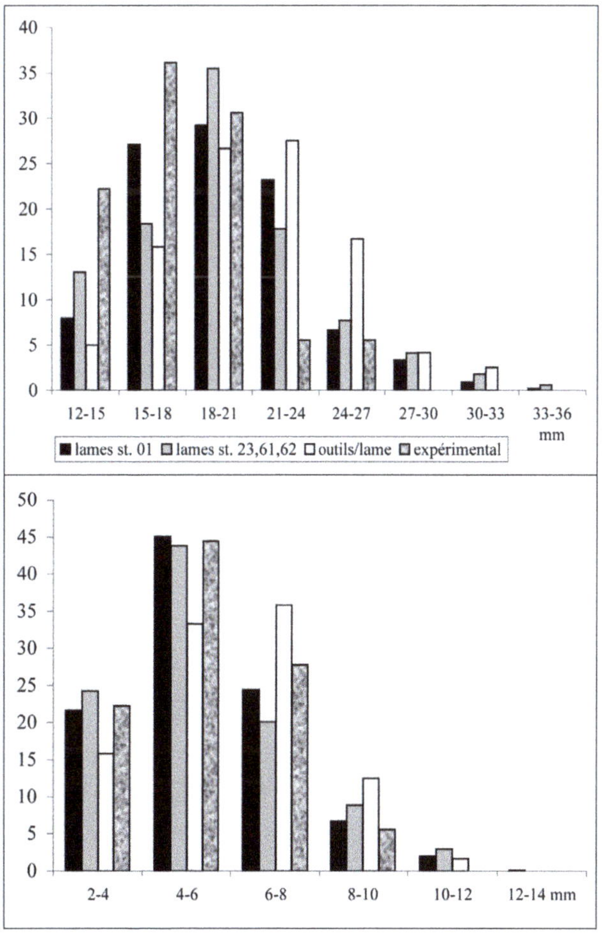

Fig. 3: Verlaine 'Petit Paradis'. Dimensions of the experimental blades and of the archaeological blades.

Theoretical reconstruction based on blade dimensions and the reduction of volumes (method 2)

The second method corresponds to a theoretical 'ideal' scheme of the exploitation of laminar volumes (for example Astruc *et al.* to be published). The known parameters are the rhythm of the final debitage of the core (reading of a diacritic diagram of the laminar surface), the average width and thickness of the products, the section of the intended products and the negative of the laminar volume. With these parameters, it is possible to redraw an exploitation scheme of blades for each core with average laminar sections. Of course, this scheme applies only to a theoretical ideal, because it represents only perfect products. It does not account for imperfections in the flint that can modify the debitage process, knapping accidents or small laminar products corresponding to the maintenance of the core, or the regularization of its sides, for example.
This method is nonetheless pertinent in the context of a standardized core reduction made by indirect percussion, such as at Verlaine.

Refits (method 3)

In addition to pieces associated with core preparation and the opening and repairing of striking platforms, whose refitting allows us to reconstruct the laminar volumes, numerous blade fragments (620) are also present in the concentrations. In two cases, most of the blades could be refit, thus allowing an exact restitution of the debitage (both cases are blades broken in the middle during debitage). These are the most pertinent refits, as they give a true vision of the range of blades produced and the small adjustments made on the sides of the core or its laminar surface.

4.3. Application

These different approaches were tested on a few cases in order to compare estimations of the number of blades produced.
Case 1: We were able to apply a triple approach to core 18, thanks to the 5 blades that could be refit onto it. The archaeological debitage shows the extraction of 14 blades. The sequence corresponds perfectly to that most often recorded on the cores: the extraction of lateral blades to prepare the central blade with three sections, in a 2-1-2' rhythm.

The most probable theoretical scheme yields a total of 16 blades, but that includes many small ones. Confrontation with the experimental debitage yields a total of 16-17 blades.
With these parameters, we thus obtain a difference of 2 to 3 blades relative to core 18, or 14% in this particular case (6% between the experimental and the theoretical scheme). This difference is clearly explained by the knapper's management of the volume, beginning the debitage with a thick blade prepared by bladelet removals, and the preparation of the left side by the removal of short bladelet-blade products (Fig. 4A).

Case 2: In this case, we tested the theoretical and experimental schemes (Fig. 4B). We obtained 23 blades for the first and 25 for the second (8.7% difference).

Case 3: Finally, a last attempt was made with core 65 (Fig. 4C), on which we were able to refit the opening blade and a cortical blade. It allowed us to rather precisely refine the theoretical scheme to 8-9 blades as opposed to 9 for the experimental debitage (12,5% difference).

5. RESULTS

In this paper, we propose only a first estimation of the production, because all three methods were not systematically applied to the assemblage of cores of concentration 130. The only exhaustive estimation conducted was that of experimental debitage. This was possible for 58 blocks out of 66, because some had no production: they had been abandoned before the extraction of standardized blades, while others were impossible to estimate because the striking platform was natural and never repaired. Finally, a few cores collected by Neolithic people themselves are missing from the total assemblage.

The 58 blocks that furnished laminar blanks correspond to 65 distinct cores. Based on the method of experimental debitage, an average of 21 to 23 blades were produced per core. The estimation of the number of blades produced thus oscillates between 1365 (65x21) and 1495 (65x23). This is a minimum interval, since for at least 5 cores we have only the second phase of exploitation, which is posterior to a core tablet. By reference to the cases described above, the difference between the experimental and theoretical scheme is roughly 1 (cases 1 and 3) to 2 (case 2) blades

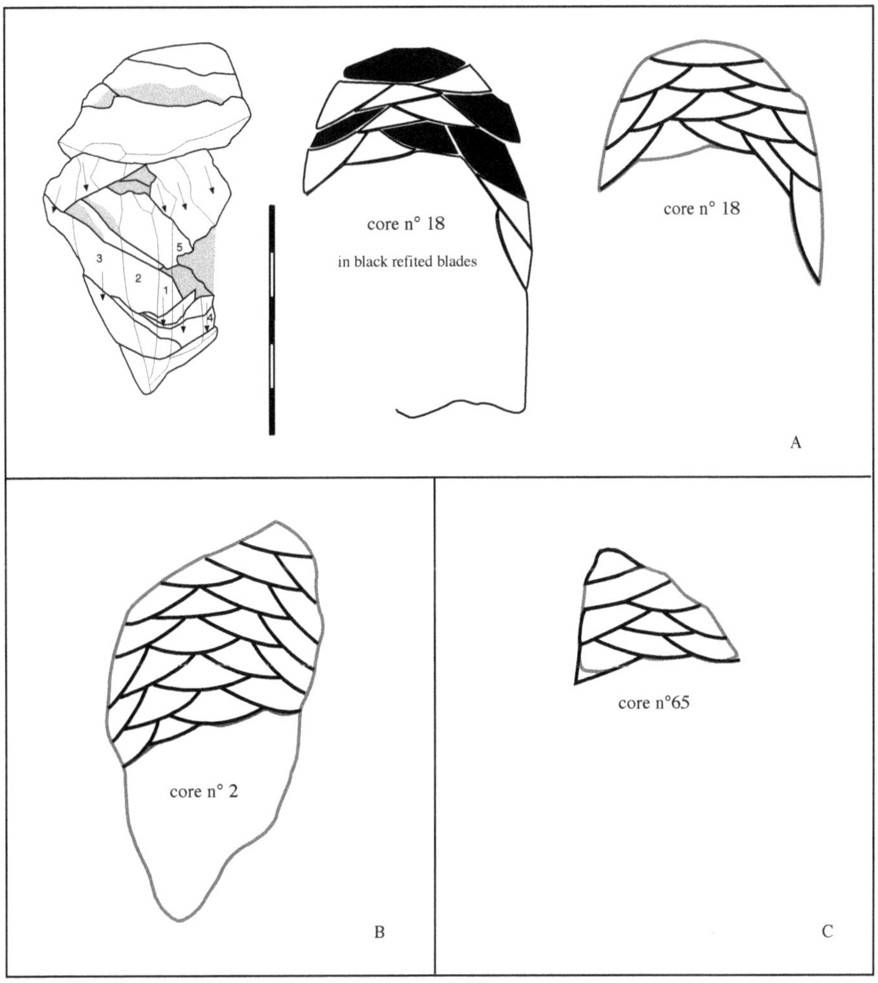

Fig. 4: Verlaine 'Petit Paradis'. The production calculation applied to 3 cores of feature 130. Restitution of laminar debitage with the refited blades (method 3) and restitution with theoretical scheme of the exploitation (method 2) n° 18, 2, 65.

per core. By extrapolation, this gives from 19 (21-2) to 21 (23-2) blades per core for the lowest range, and from 20 to a maximum of 22 specimens, for a total number of blades within the range of 1235 (19 x 65) /1365 (21 x 65) to 1300 (20 x 65) /1430 (22 x 65).

Finally, by comparing with the only archaeological reference in this example, the difference would be from 2 to 3 blades, which would give 18 to 20 blades per core, and thus a total number of 1170 to 1300 specimens.

These three estimations are in fact pretty close, and a range between 1200 and 1500 blades seems correct when we consider the three possibilities, or between 18 and 23 blades per core. This range is approximate but gives an idea of what can be done by confronting methods and data.
The concentration contained 620 whole or fragmented blades. A calculation of the minimum number of items was attempted by counting the whole and almost whole blades and the distal parts (the most numerous), which corresponds to 327 products.
Therefore, in concentration 130 at Verlaine, for the production of the 58 blocks accounted for, we suggest that between 900 and 1200 blades were collected by the Neolithic inhabitants.

The relationship between the quantity produced and the quantity used *in situ*

The problem of the quantification of products consumed in spatial units of the village can be only partially treated here, as the study of artefacts from the site is still in progress, and the methods of calculation are currently being developed.
Nonetheless, the stratigraphic position of the concentration of structure 130 offers a good opportunity to study

this production within a domestic unit. The assemblage was located at the bottom of the structure, with no relation to the upper detritic level of the pit lateral to the habitation unit. We can legitimately suggest the hypothesis that the concentration resulted from the fabrication of a stock of blades for the inhabitants of this house. We thus analyzed the objects in the domestic pit made from the fine-grained Hesbaye flint and compared them with those of the assemblage.

This second assemblage differs in that it contains tools (including cores reused as hammerstones), used flakes and burned objects, artefacts already observed in other similar structures in this site.

The artefact assemblage of the detritic level also includes the ensemble of laminar debitage waste products, in proportions comparable to that of the concentration. The only real difference lies in the number of bladelets and laminar flakes, but this could be related to the archaeological technique (the concentration was removed in one block and sifted). Elements of all stages of laminar debitage are present, from unworked blocks to cores. There are 18 cores, including three reused as hammerstones and three fragments. Retaining the estimation proposed based on the concentration, we obtain a range from 378 (18 x 21) to 483 (23 x 21) blades for 21 cores, or, if we consider only the pieces not reused as hammerstones, from 144 (18 x 8) to 184 (23 x 8) blades.

There are 154 unretouched and retouched blades in the detritic level. This number is thus equivalent, or even inferior to the number estimated based on the debitage waste of this assemblage.

CONCLUSION

The test that we have proposed is based on the integration of three methods applied to the artefacts of a debitage concentration and of detritic waste found in the same structure. The estimation of the average minimum number of blades produced per core is 18 to 23, resulting in a minimum of 1200 to 1500 blades for the entire pit. In contrast, the unretouched or retouched blanks of this assemblage, or blades used *intra-muros*, represent approximately 40% of the blades in the detritic levels and 10% to 20% relative to the minimum number of blades produced based on the debitage waste of the entire pit. This large deficit compels us to further quantitative analyses in the entire house and ultimately the site. We will also analyze the qualitative variability of the blades produced and used. Finally, we must consider the entire functional spectrum, to reveal the relative proportion of the exploitation of the sub-products of laminar debitage. These parameters enable us to better define the dichotomy between an overabundant production whose objective is not a functional *intra-muros* exploitation, but export; simultaneously, a weak *in situ* consumption of products corresponding to a reduction sequence identical to the preceding one, applied to the same raw material. Integrating these different aspects will allow us to better conceptualise the complexity of the objectives which informed the management of the production of tool blanks in the Bandkeramik of Hesbaye.

Pierre Allard
CNRS-UMR 7055 'Préhistoire et Technologie'
21 allée de l'Université
F-92023 Nanterre Cedex
e-mail: pierre.allard@mae.u-paris10.fr

Laurence Burnez-Lanotte
University of Namur FUNDP
Laboratoire de Protohistoire européenne
61 rue de Bruxelles
B-5000 Namur Belgique
Associée UMR 7041 CNRS
e-mail: laurence.burnez@fundp.ac.be

REFERENCES

Allard, P. 2005. *L'industrie lithique des populations rubanées du nord-est de la France et de la Belgique.* Internationale Archäologie, 86. Rahden/Westphal: Verlag Marie Leidorf GmbH.

Allard, P. 2007. Surplus production of flint blades in the early Neolithic of Western Europe: new evidence from Belgium. *European Journal of Archaeology* 8, 2005, 205-223.

Allard, P. and Burnez-Lanotte, L. 2007. Surplus production in the Belgian Lincarbandkeramik: blade debitage at Harduémont 'Petit Paradis' (Verlaine, Hesbaye, Belgium).

In *Stone Age-Mining Age, VIII International Flint Symposium. Bochum, September 1999*, Der Anschnitt beiheft 19. Bochum, 37-54.

Astruc, L., Gartuze, B., Pelegrin, J. and Akkermans, P. in press. From Production to use: a parcel of obsidian bladelets at Sabi Abyad II. In L. Astruc, D. Binder and F. Briois (eds.), *La diversité des systèmes techniques des communautés du Néolithique pré-céramique: vers la caractérisation des comportements sociaux, 5ème colloque international sur les industries lithiques du Néolithique pré-céramique*. Antibes: Éditions APDCA 2007.

Binder, D. and Gassin, B. 1988. Le débitage laminaire chasséen après chauffe : technologie et traces d'utilisation. In S. Beyries (ed.), *Industries lithiques, tracéologie et technologie*. B. A. R., Int. Ser., 411, Oxford : BAR, 93-125.

Burnez-Lanotte, L. and Allard, P. 1998. Mode de production laminaire dans le Rubané de Hesbaye au 'Petit Paradis' à Harduémont (Verlaine, Lg). In *Organisation néolithique de l'espace en Europe du nord-ouest, XXIIIème colloque interrégional sur le Néolithique, 24-26 octobre 1997 à Bruxelles*, Anthropologie et Préhistoire, 109, 15-26.

Burnez-Lanotte, L. and Allard, P. 2003. Blade debitage in the Belgium Linearbandkeramik: the production at Harduémont 'Petit Paradis' (Verlaine). In L. Burnez-Lanotte (ed.) *Production and Management of Lithic Materials in the European Linearbandkeramik, UISPP, Septembre 2001, Liège, Belgique*, B.A.R. Int. Ser. 1200, Oxford: BAR, 59-64.

ACKNOWLEDGEMENTS

We are very grateful to the Fyssen Foundation for the interest shown in our lithic research and for funding Pierre Allard a post-doctoral year at Namur. We would also like to thank Jacques Pelegrin (CNRS-UMR 7055) for his comments and his help about the theoretical 'idea' scheme of the exploitation of laminar volumes.

THE PREHISTORIC FLINT MINING COMPLEX AT SPIENNES (BELGIUM) ON THE OCCASION OF ITS DISCOVERY 140 YEARS AGO

Hélène Collet, Anne Hauzeur, Jacek Lech

Abstract: 140 years ago, the Neolithic flint mining complex at Spiennes, Hainaut province, became known to Europe and the world. That discovery spurred the dynamic development of research into prehistoric flint mining in European and world archaeology. This anniversary is an occasion to sum up what we know today about this mining complex which, in the year 2000, was placed on the UNESCO world heritage list.
Excavations of the Spiennes flint mine sites have been a long and rich tradition producing important results for European prehistory. These results are presented in the first part of the paper. In the second part, the authors sum up the current state of knowledge about the mining fields at Spiennes, about the various methods of flint exploitation, production trends and the chronology of mining activities. Particular attention has been paid to studies of old chipped inventories and to the most significant results of recent research led by H. Collet.

Keywords: Spiennes, Neolithic, prehistoric flint mining, methods of chipping floor research, flint cores and blades, flint axe, Michelsberg culture, history of archaeology.

INTRODUCTION

Spiennes is one of the most important names on the map of prehistoric flint mining. Today, the village lies on the outskirts of Mons, the capital of the Hainaut province in the region of Wallonia, Belgium (Fig. 1). For ages, the place has been known for the vast number of flints lying on the surface of the fields. Towards the end of the first half of the 19th century, collectors of antiquities became more and more interested in these flint pieces. While Spiennes' fame was not equal to that of Grand-Pressigny in France, it was visited by Belgian and foreign antiquarians who collected ancient flint artefacts.
140 years ago, in 1867, a spectacular archaeological discovery was made at Spiennes. A railway cutting revealed the deep shafts of a Neolithic flint mine and its underground galleries. Spiennes became known to the world. The 1867 find was followed by others, so that in the second half of the 19th century and the first decades of the 20th, the Spiennes micro-region became famous as a place where many important discoveries connected with prehistoric flint mining had been made, influencing developments in the archaeology of prehistoric flint mining in Europe and North America. When discussing the places where flint tools were made, John Evans (1872: 29-32) compared Spiennes to Grimes Graves and Cissbury in England, and to Grand-Pressigny in continental Europe. His views were in accordance with those of Albert Toilliez, who first studied Spiennes. Today, the prehistoric mining complex at Spiennes is considered one of the most important features of prehistoric flint mining in the world, next to Grimes Graves and Cissbury in England, Rijckholt-St Geertruid in the Dutch Limburg, and the Krzemionki Opatowskie in Poland. It has been on the UNESCO World Heritage list since November 30, 2000.

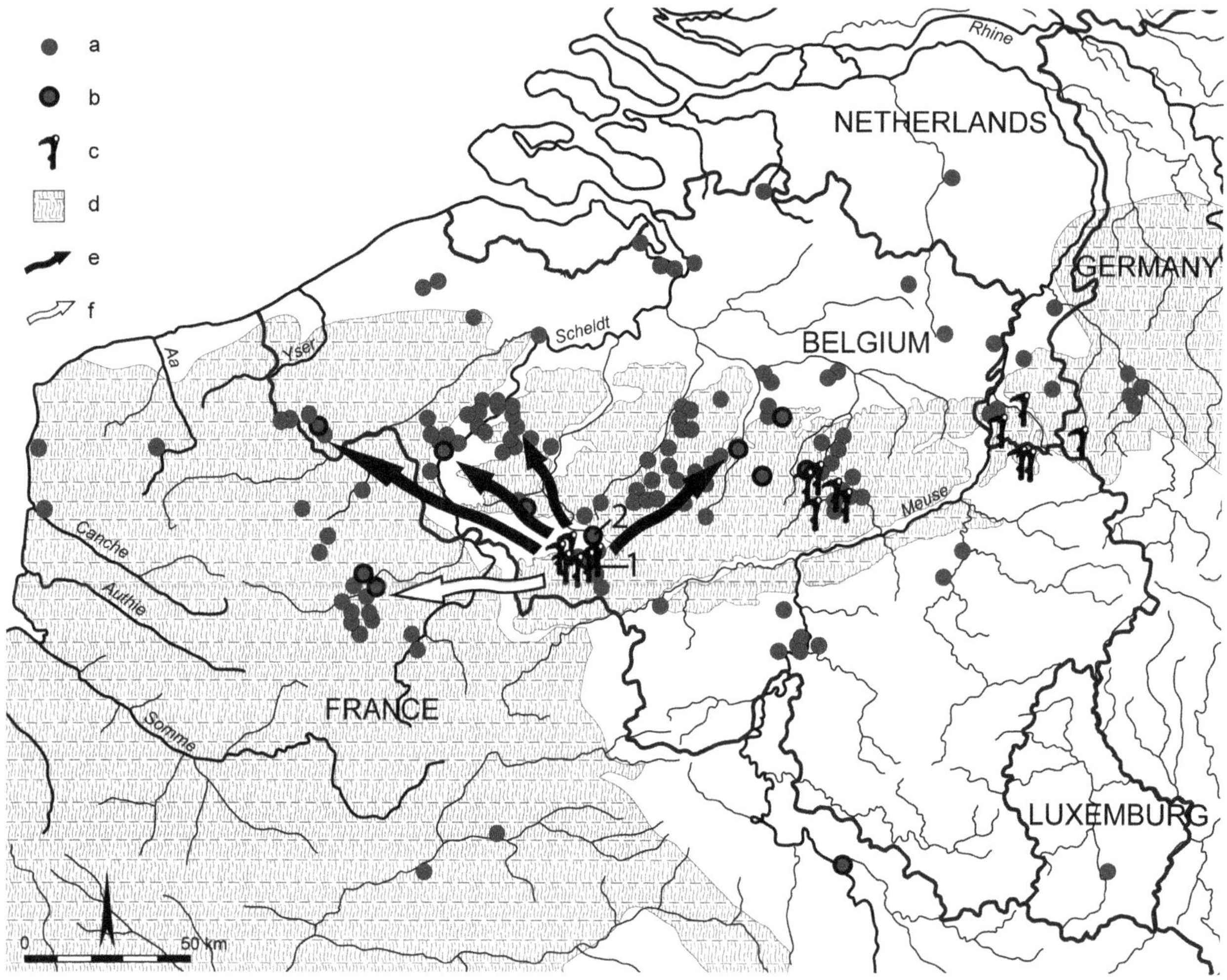

Fig. 1: Location of the prehistoric mining complex at Spiennes, Hainaut province (Belgium) and other important archaeological sites. 1. Spiennes flint mine. 2. Thieusies. a - Middle Neolithic settlements and flint concentrations; b - Middle Neolithic enclosure sites; c - flint mines; d - loess area; e - Spiennes flint distribution; f - supposed Spiennes flint distribution. After Vanmontfort *et al.* (in print).

1. CONTEXT OF THE MINING FIELD AT SPIENNES

1.1. Geography and geology

The mining field at Spiennes is located in the district of Mons, about 5 km south-east of the city of Mons, in the Belgian Province of Hainaut. The centre is situated 123,26 N and 123,19 E by the Belgian Lambert coordinate system and 50° 25' 11,01"N and 3° 59' 29,99" E by the geographical coordinate system WGS84.

From a geographical and geological point of view (Fig. 2), the mining field belongs to the Mons basin, which is a west-east oriented sink where the river Haine flows, a tributary of the river Scheldt. To the north, it is surrounded by the low plateaus of Hainaut, that of Anderlues to the east and the Haut-Pays to the south. The mining site is located on the northern flank of the Harmignies cuesta, a crest with asymmetric slopes extending from east to west, which culminates at a height of 93 m. The cuesta is cut by the river La Trouille and the river La Wampe, tributaries on the left side of the river Haine. The substratum of this cuesta consists of Upper Cretaceous chalk covered by Tertiary glauconitic sands and Wechselian loess. From top to bottom, the chalk 'formations' are Craie de Spiennes, Craie de Nouvelles, and Craie d'Obourg. The Spiennes 'formation', of which the basal 10-12 m date from the upper part of the Upper Campanian (Robaszynski, Christensen 1989), includes many thick black flint seams. The Neolithic miners exploited several of these.

1.2. The mining field

The site is located in fields south of the village of Spiennes (Hainaut, Belgium) occupying the slopes of La Trouille valley, parts of the two plateaus surrounding it and the slope toward the river La Wampe (Fig. 3). Flint mines and

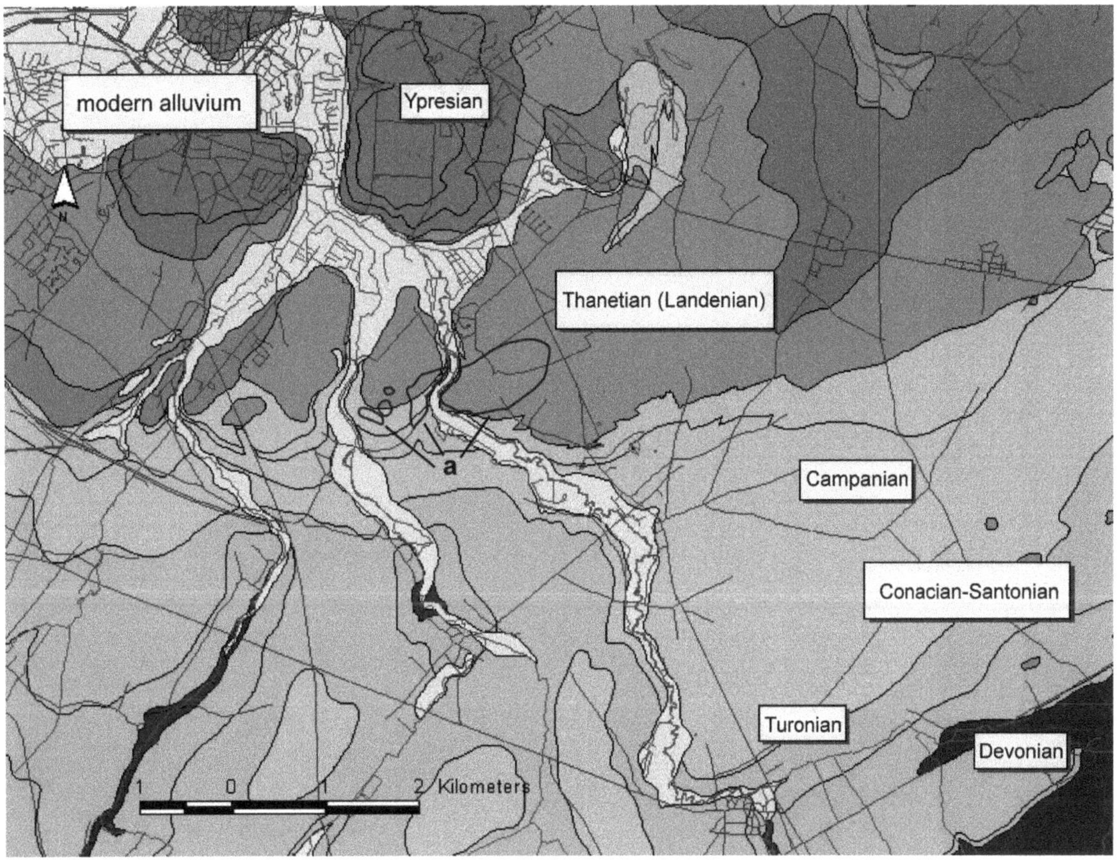

Fig. 2: Geological map of the Spiennes flint mine. Design and drawing by O. Collette. a – extent of flint mine based on surface mined flint distribution plotted by F. Gosselin.

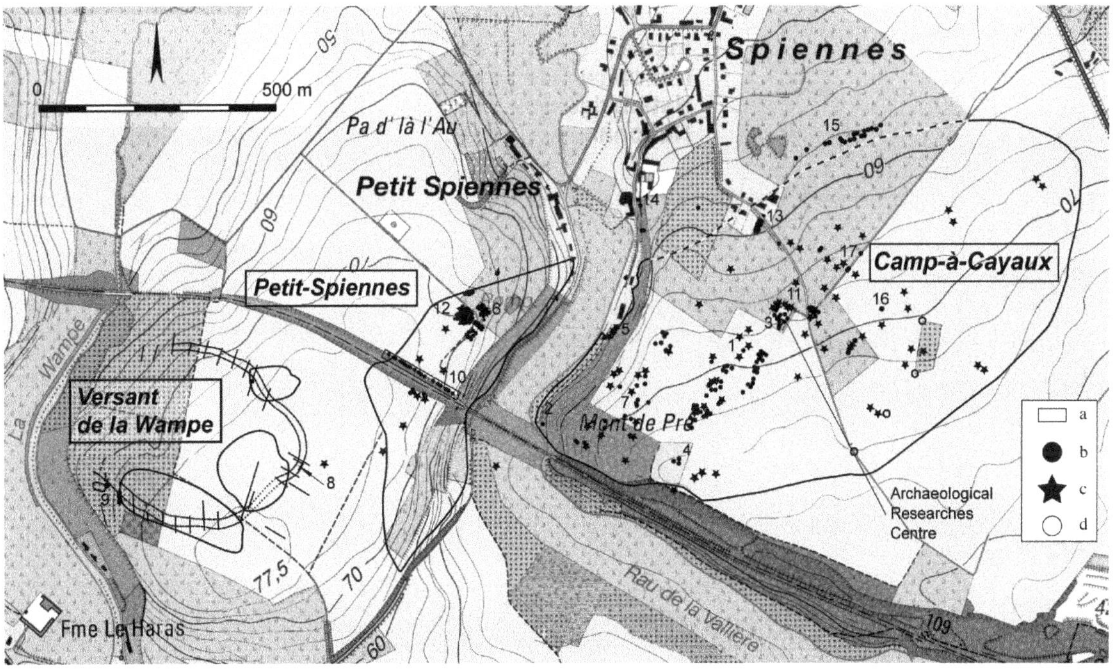

Fig. 3: The prehistoric mining complex at Spiennes. Map of the site with the most important places excavated so far.
a - 1867 railway trench; b - shafts; c - workshops; d - geological anomalies; 1 - 1925 A. de Loë and E. Rahir excavation; 2 - approximate location of a gallery found by L. De Pauw and E. Van Overloop in 1889-1890; 3 - 1912-1914 A. de Loë excavation of deep shafts 1 and 2; 4 - approximate location of 1930 J. Breuer excavation; 5 - 1953 J. Verheyleweghen excavation in the slope of the La Trouille river valley; 6 - SRPH excavation since 1953; 7 - 1965 F. Hubert excavation; 8 - interruption of the Michelsberg enclosure; 9 - shafts anterior to the Michelsberg enclosure; 10 - 1975 F. Hubert excavation of shafts and mining pits near the railway trench; 11 - 1990 F. Hubert and M. Soumoy excavation of shallow shafts; 12 - 1997-2004 H. Collet excavation at *Petit-Spiennes*; 13 - 2005 H. Collet excavation of chipping floors on plot 51c; 14 - 2005 H. Collet excavation in the slope of the La Trouille river valley; 15 - shafts discovered by the scientific society *Recherches et prospections archéologiques en Wallonie* during the laying of a gas pipeline; 16 - 1962 J. Verheyleweghen excavation of a shaft; 17 - evidence of shafts discovered in 2004 by H. Collet on plot 169a east of Harmignies Street.

workshops cover about 100 hectares (Hubert 1980: 124). The mining site can be divided into three different mining areas. The largest one, called *Camp-à-Cayaux* (about 65 ha), is located on the plateau east of the river La Trouille. The second, known as *Petit-Spiennes* (about 14 ha), is situated on the plateau west of La Trouille and extends parallel to the river. A third one, named *Versant de La Wampe*, discovered later, is located on the plateau of *Petit-Spiennes* in the part surrounding the river La Wampe. This is also where an enclosure was built during the Neolithic period.

2. HISTORICAL BACKGROUND

2.1. The first discovery by Albert Toilliez

Albert Toilliez (1816-1865), a mining engineer in Mons, was the first important researcher at the Spiennes archaeological site (Fig. 4).

Before this 1850's discovery, some shafts and galleries, most probably Neolithic, had been observed for the first time in the summer of 1842, near the southern and western rim of the plateau of *Camp-à-Cayaux*, at a place called *Mont de Prêle*. These extracting features were mentioned in local newspapers at the time, but their date was unknown.

Fig. 4: Mons, Hainaut province. Grave of Albert Toilliez (1816-1865) the first important researcher of the Spiennes flint mining complex. Photograph by A. Hauzeur.

As early as 1851, Albert Toilliez suspected the probable existence of an axe workshop at Spiennes. His cousin Désiré Toilliez reported that Toilliez had collected more than 60 flint axes, and roughouts, in the district of Harmignies and in the district of Spiennes, where these artefacts may have been manufactured (Toilliez 1851).

Shortly after, a date later than Palaeolithic was proposed for these workshops, despite the fact that some researchers at that time held to the older chronology. This conviction was not based on stratigraphic evidence but only on similar morphology between hand axes and axes roughouts from Spiennes (De Koninck 1860).

After A. Toilliez's death in 1865, his collection seems to have been bought by the famous British antiquarian, field archaeologist and numismatist Sir John Evans (Hamal-Nandrin, Servais 1925: 75), and then sent to Oxford.

2.2. The railway trench in 1867

In 1867, the building of the railway from Mons to Chimay required the digging of a deep trench to cross the plateau of *Petit-Spiennes* (Fig. 5). Thus were discovered several spectacular cross sections of flint mine shafts in the wall of the trench.

In 1868, the controversy over the chronological position of the Spiennes' artefacts came to an end with the publication of the report *'sur les découvertes géologiques et archéologiques faites à Spiennes'* (Briart *et al.* 1868). The authors claimed that there had been two periods of human occupation at Spiennes, with the extensive workshop and the digging of shafts occurring during the later period. The numerous chipped flakes found in the surroundings of Spiennes had been made in situ with materials extracted in the vicinity.

The dating of the shafts was determined on the basis of polished axes, potsherds tempered with flint, and bones of recent fauna, including those of the domesticated dog. This referred to the workshops of Grand-Pressigny, and to the fact that a dagger was discovered at the surface at Spiennes. At that time, this artefact was thought to have

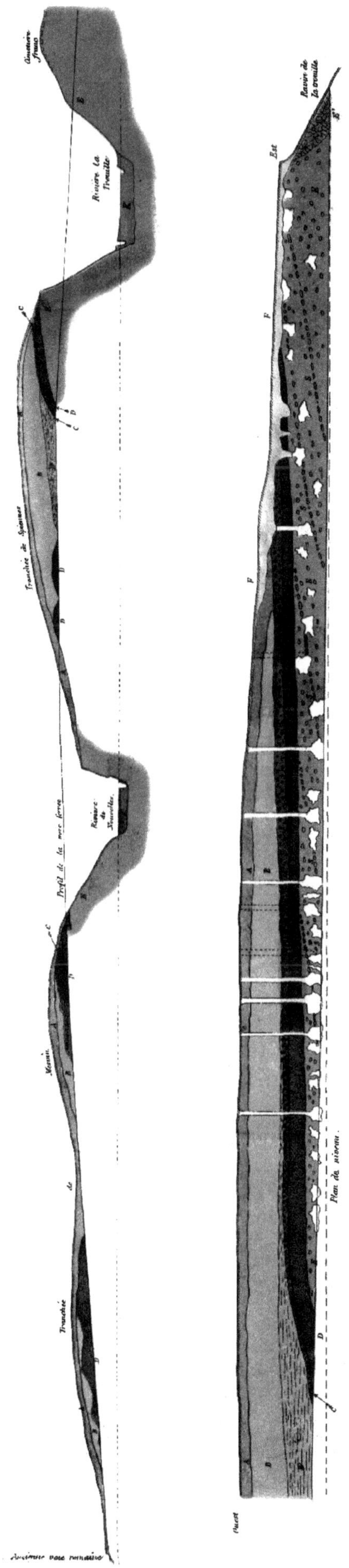

Fig. 5: Spiennes, Hainaut province. *Petit-Spiennes*. Cross section of the 1867 railway trench. After A. Briart, F.-L. Cornet and A. Houzeau de Lehaie (1868).

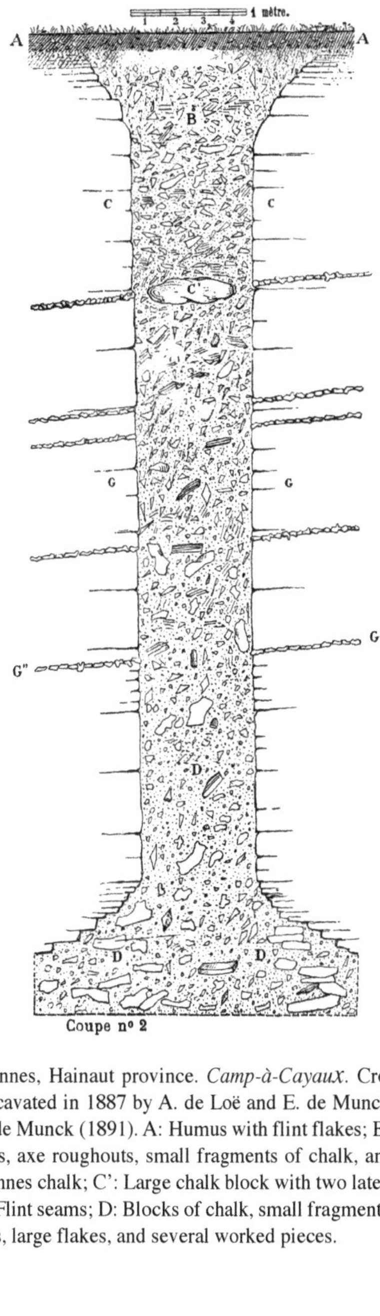

Fig. 6: Spiennes, Hainaut province. *Camp-à-Cayaux*. Cross section of the shaft excavated in 1887 by A. de Loë and E. de Munck. After A. de Loë and E. de Munck (1891). A: Humus with flint flakes; B: Flint flakes, cores, blades, axe roughouts, small fragments of chalk, and chalk powder; C: Spiennes chalk; C': Large chalk block with two lateral grooves of a rope; C": Flint seams; D: Blocks of chalk, small fragments and powder, flint nodules, large flakes, and several worked pieces.

been made of Grand-Pressigny flint[1]. For the authors, this demonstrated exchanges between both workshops (Briart *et al.* 1868: 379).

The results of the excavations were presented by Alphonse Briart (1825-1898), François-Léopold Cornet (1834-1887), and Auguste Houzeau de Lehaie (1832-1922) at the International Congress of Anthropology and prehistoric Archaeology held in Brussels in 1872. Publication in the proceedings gave the study and the site of Spiennes some celebrity.

1. Today this piece is surmised to be the dagger from the Houzeau de Lehaie's collection. According to M. Delcourt-Vlaeminck, it was not made of Grand-Pressigny flint (pers. comm.).

In spite of its importance, the discovery was not followed up by a serious programme of research and excavation. The only subsequent investigations were the excavation of a mine shaft - not the galleries - undertaken in 1887 by Baron Alfred de Loë (1858-1947) and Emile de Munck (1861-1944). Those excavations are situated in the heart of *Camp-à-Cayaux* to the south west of the present *Centre de Recherches archéologiques*. The first, schematic, cross section of the shaft excavated shows that the flint seam exploited appears to be the sixth (Fig. 6). Moreover, excavations explored some upper parts of shafts, workshops and other pits. The results were presented at the International Congress of Prehistoric Anthropology and Archaeology held in Paris in 1889 (Loë, Munck 1891).

This was also the period of the first technological studies of the material, such as a suggested method of blade production by indirect percussion. These observations were supported by the discovery of bone punches (Cels, De Pauw 1885-1886). Louis De Pauw and Eugène Van Overloop (1889-1890) noted, for the first time, a gallery dug directly into the slope of the valley, below the western part of *Camp-à-Cayaux*.

2.3. The excavations of Baron Alfred de Loë

The excavation of two shafts at *Camp-à-Cayaux* by the National Excavation Service, based at the *Musées du Cinquantenaire*, began in July 1912 thanks to a grant from Count Louis Cavens. These truly ambitious excavations were conducted by Baron Alfred de Loë and continued in 1913 and 1914 (Loë 1925). Two shafts, 16 m deep, were excavated in 1912 (Loë 1925; Loë, Rahir 1929). Parts of the galleries were explored in 1913, as well as some workshops in the vicinity of the shafts. More than 1,500 picks were collected during these campaigns, as well as several hammers in Landenian or Devonian sandstone. Thin potsherds tempered with very fine flint chips were collected. They are part of the 'chalice-shaped' pots, attributed by François Hubert to the Michelsberg Culture (Hubert 1978).

Pictures of the galleries and simplified cross sections of the two shafts, with the drawing of some layers, were published. Unfortunately, only general information was given about the archaeological material (for example, Loë 1925).
More galleries were later explored by Maurice Lefort, in excavations headed by Jacques Breuer around the 1950's. An area of about 250 m² was investigated. A very detailed and exhaustive study of these shafts was conducted by F. Hubert in 1978, with a detailed underground map, a fine stratigraphy of the geological background, and a description of the picks (Hubert 1978). New excavations were conducted in 1982 by F. Hubert (1983) who reported on his findings but did not offer a detailed study of the archaeological material.

2.4. Field research

Baron Alfred de Loë and Edmond Rahir also excavated in 1925 some 34 'sunken-floor huts' as they called them (Loë, Rahir 1929); actually, these should be considered heads of shafts. Similar surface explorations of the site were conducted in 1924 by Jean Hamal-Nandrin, professor at the State University of Liège, and Jean Servais, curator of the Archaeological Museum of Liège (Hamal-Nandrin, Servais 1925). Others were made by Charles Stevens in the beginning of the 1920's for Aimé Rutot, curator of the Museum of Natural History in Brussels. Charles Stevens, a local quarry worker, was active over a very long period (more or less between 1911 and 1953), and excavated on his own or on request. He himself chose the places to excavate. As he left no record of where exactly he had been digging, nor any drawings, we have hardly any information about the features he excavated, neither their morphology nor the nature of the fillings. His main purpose was to enlarge the collections of museums and collectors, with no concern for such details as contexts. His research has been published only partially, if at all.

In 1930, Jacques Breuer (1892-1971) applied new and improved techniques of excavation. He dug real archaeological test trenches to systematically determine the kind of structures. Breuer identified features such as temporary open air mining zones. He drew attention to the scarcity of artefacts, especially pottery and domestic remains (Breuer 1930).

2.5. Jean Verheyleweghen's excavation and interpretation

This tradition of surface test trenches was also practised by Jean Verheyleweghen (1910-1965) who excavated nearly 250 features between 1945 and 1958. These were interpreted as knapping and chipping places, funnels of shafts, some occasional pits of domestic waste, and one secondary burial. The structures were widespread over the whole surface of *Camp-à-Cayaux*, and along the slope of the right bank of river La Trouille (Verheyleweghen 1963).

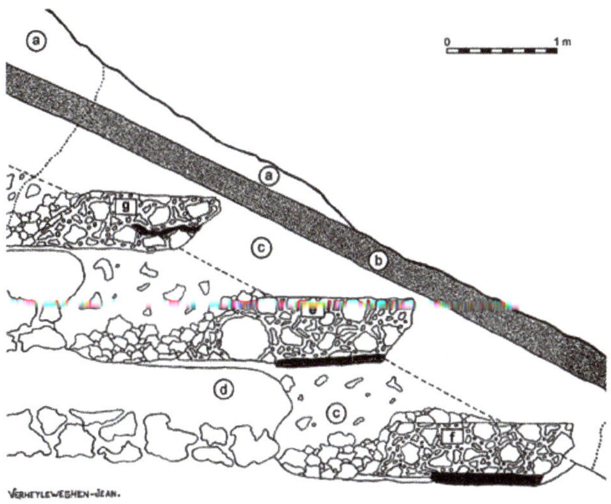

Fig. 7: Spiennes, Hainaut province. *Camp-à-Cayaux*. The slope of the La Trouille river valley. Cross section of flint exploitation remains found in 1953 by J. Verheylewhegen: a - disturbed sediment; b - humus; c - yellow sediment mixed with chalk in blocks and powder; d - natural chalk rock and flint seams; e, f, g - 'workshops' according to J. Verheyleweghen. After J. Verheyleweghen (1953).

These important field campaigns allowed him to propose an outline of the development of the mining field at *Camp-à-Cayaux*. This schematic evolution is divided into four periods. The first three developed from a simple excavation along the slope of the valley to the complex mining works of the deepest shafts. The fourth phase corresponds to the decline of the mining activities. The outline was taken as a model to interpret the flint mine site of Rijckholt St-Geertruid (NL), and Cissbury (GB). Despite its importance, this work has never yet been critically discussed, mostly because of the lack of later large-scale excavations and accurate chronology. Verheyleweghen's results and ideas were fully accepted by Siegfried J. De Laet, and published in his book *La Belgique d'avant les Romains* (1982).

2.6. The beginnings at Petit-Spiennes

The first excavations of the *Petit-Spiennes* plateau began in 1953. They were organised by several members of the society *Les Amis du Musée de Mons* which later became known as the *Société de recherche préhistorique en Hainaut*. The localisation of the excavation site was chosen by André Adam (1919-2005) who had noticed concentrations of flint flakes, waste, and fragments of chalk on the ground surface. A first shaft and the upper part of a second one were excavated at this site, until 1960. After a pause, research started again in 1979, at the same site exploring other shafts and underground galleries. Work here is still in progress, with 95 m² so far excavated.

Marcel Lefrancq (1916-1974) with the help of Paul-Henri Moisin (1927-2004), both members of the Society, led the excavations, supported by professionals archaeologists. Their work and publications were of excellent scientific quality, even if we deplore the absence of drawings of cross sections.

Between 1988 and 1992, excavations were carried out by Jean-Pierre Joris with the help of Françoise Gosselin, who published the first monograph of the *Petit-Spiennes* area, including all the results since the beginning of excavations in 1953 until the date of publication (Gosselin 1986).

2.7. François Hubert's research and excavations

The next important step in the research at Spiennes was done by François Hubert, assistant in the *Service national des Fouilles* at that time. He undertook excavations both in the *Camp-à-Cayaux* area and on the *Petit-Spiennes* plateau.

In 1965, François Hubert excavated four workshops as well as the upper part of three shafts at *Camp-à-Cayaux*. His finds included some Michelsberg pottery (Hubert 1969). A field survey allowed him to determine more precisely the extent of the mining field. In 1975, he conducted a rescue excavation initiated by the construction of a gas pipeline crossing the railway at *Petit-Spiennes* (Hubert 1976a). All these excavations were highly professional and provided interesting data about material, stratigraphy and methods of exploitation.

Since January 1954, André Adam, one of the founders of the *Société de recherche préhistorique en Hainaut*, had suspected that a settlement existed on the plateau of *Petit-Spiennes* south of the railway. Indeed a Michelsberg enclosure was in fact discovered right at that spot (Adam 1959). The same year, Irwin Scollar (1955) noticed the presence of crop-marked fragments of two concentric circles when he examined aerial views. Between 1966 and 1979, F. Hubert carried out annual excavations in the ditches, by making trenches at regular intervals. Only the first years of his investigations have been published (Hubert 1971), as well as the ceramic vessels (Vanmontfort 2004).

During the annual excavation of the ditches of the fortified camp in 1970 and 1974-1975, flint extraction features (shafts and gallery roof), earlier than the ditches, were discovered (Hubert 1971: 50-51; Hubert 1976b: 12-15). The inner area of the fortified camp is still almost completely unexplored. Only one trench 2 m wide and 108 m long was dug inside the camp.

3. THE LATEST DECADES OF RESEARCH AT *PETIT-SPIENNES* AND *CAMP-À-CAYAUX*

Several kinds of explorations took place at *Camp-à-Cayaux* during the 1990's, giving us a better idea of the extension of the mining area. Preventive excavations around the *Centre de recherches archéologiques* yielded five shallow shafts (Hubert, Soumoy 1993; Bonenfant, 1993). Rescue excavations, prior to the laying of a gas pipeline, allowed for the detection of 20 more shafts located in the northern part of the mining site, and contributed to the extension of the known mining area in this direction. A small Neolithic workshop was excavated in the south-western part of the mining site, in the valley of La Trouille (Fechner *et al.* 1993).

Preventive excavations were also undertaken at the mining site of *Petit-Spiennes*, because there was a project to develop the archaeological site (Collet *et al.* 1997). Luckily the project was put aside, and the excavations turned into planned ones, led by Hélène Collet. Since there is no emergency in this area, and rescue work is needed much more urgently elsewhere, these excavations are still continuing.

Several features were identified, among which some mine shafts should be noted. A deep shaft was fully excavated, as well as half of another; there was also what was interpreted as an aborted shaft, as well as shallow pits. So far only provisional papers have been published regarding this research (Collet 2000; 2003b; Collet, Vander Linden 1998; Collet, Van Neer 2002; Collet, Woodbury 2000; 2001; 2002; Collet, Vankerkhoven 2004), but both features and material are being studied. At the same time the area investigated by the volunteers of the SRPH since 1953 continues to be explored. A new mine connected with the previous ones has been excavated and the filling of two shafts is currently being explored (Desterbecq 2004; Desterbecq, Joris 2006; Robert 2003).

3.1. Rescue excavations at *Petit-Spiennes* and research results

In the case of systematic excavations, it was decided to record the data as fully and carefully as possible. For this reason, the choice was made to draw very detailed cross sections of the filling of the shafts. In the case of rescue excavations, a strategy of recording of arbitrary 20 cm levels was adopted; later however, this was replaced by recording the real observed stratigraphic layers. Detailed stratigraphy of the extraction features made it possible to better establish the different stages of filling and to try to calculate the speed of the filling process. To confirm or to help understand this question of deposit speed, stratigraphic data were crossed with the paleoenvironmental data: palynological, archaeozoological (primarily microfauna and snail remains), and anthracological (Collet, Van Neer 2002; Defgnée, Collet 2003).

3.2. The discovery of a chipping floor at *Camp-à-Cayaux*

For some years now, rescue and survey excavations have been conducted in the surrounding area, in the vicinity of the village of Spiennes, prior to the building of a fence, house or renovation of the garden of the *Centre de recherches archéologiques*.
In early 2005, the archaeological team undertook a rescue excavation, prior to the building of a house, on lot 51c which lies just outside the part of the site under Cultural Heritage protection, but inside the archaeological site of *Camp-à-Cayaux*. During this investigation, 17 extraction features were discovered over an area of about 150 m^2, as well as remains of *in situ* workshops.

4. SPIENNES AND RESEARCH CONCERNING PREHISTORIC FLINT MINING IN EUROPE

The Spiennes mine offers one of the longest histories of archaeological excavations in Europe, indeed the world, and has played a major role in research into prehistoric flint mining. It was presented at International Congresses of Anthropology and Prehistoric Archaeology in the 19th and early 20th century, and has always been discussed in archaeological textbooks (e.g. Mortillet 1885; Cartailhac 1896; Déchelette 1908; Childe 1925). Especially important was the famous Spiennes Section of a railway cutting, discovered in 1867, still a reference even in the 20th century, although archaeological methods and our knowledge of prehistory have progressed enormously since then (Clark, Piggott 1933; Clark 1952: Fig. 99; Piggott 1954: 39).

The discovery made by workmen engaged in railway construction in Spiennes had a direct influence on the correct interpretation of pits noted in the area of the English-hill fort at Cissbury, W. Greenwell's excavations at Grimes Graves, and on the slightly later excavations at Rijckholt-St. Geertruid (Lech 1991: 557; 1997: 611-613). Also, in later years, the Spiennes excavations inspired systematic explorations of other mines. Work at Grimes Graves was undertaken in the spring of 1914, after the spectacular results of Baron A. de Loë's excavations were made public (Loë 1913; 1914; Clarke 1915). In 1933 the Spiennes excavations were an important factor in dating the British flint mines (Clark, Piggott 1933).

With time, the systematic excavations at Grimes Graves came to play an increasingly important role in the characterizations of prehistoric flint mining in Western Europe. In *Prehistoric Europe. The Economic Basis* by Grahame Clark (1952: 174-178) Grimes Graves served as the basis for a description of prehistoric flint mining in Europe, though Spiennes continued to be much in evidence. For Gordon Childe (1957: 293), in his last edition of *The Dawn of European Civilization*, the Spiennes mine was an example of specialization of labour among the clans of Michelsberg villagers:

> 'Secondary industry and trade played a recognizable part in the Michelsberg economy. Thus at Spiennes in Belgium lived a community of specialized flint-miners skilled at sinking shafts and digging out subterranean galleries. Indeed, the Michelsberg settlers there constituted a specialized industrial community, supplementing their livelihood by exporting the products of their mines and workshops - and Spiennes was no isolated phenomenon within the Western complex. It implies also the development of hunting expeditions and transhumance into something like regular commerce.'

In *History of Mankind: cultural and scientific development*, published under the auspices of UNESCO, in the very brief description of mining and trade during the Neolithic, only Spiennes, Grimes Graves and Grand-Pressigny are mentioned (Hawkes 1965: 431-433), while in *World prehistory in new perspectives* by G. Clark (1977: 134) we find only one mention of Grimes Graves.

During the second half of the 20th century, Grimes Graves gradually took over the place of Spiennes in the reflections on prehistoric flint mining in Europe. One of the reasons for this change was the publication, in the 1920s and 1930s, of A.L. Armstrong's new findings concerning Grimes Graves and, in the 1970s, the new, extensive and very professional excavations of the site (e.g. Armstrong 1923; 1927; 1934; Mercer 1981; Longworth *et al.* 1991; Longworth, Varndell 1996). The stronger institutional position of British prehistory when compared to Belgian and its dominance over the field of archaeological syntheses also played a role. At the same time, progress was being made with excavations of other important sites of prehistoric mining, such as the ones at Krzemionki Opatowskie, Rijckholt-St. Geertruid, Defensola 'A' and Jablines (Weisgerber *et al.* 1980; Borkowski *et al.* 1991; Grooth 1991; Bostyn, Lanchon 1992; Di Lernia *et al.* 1995; Felder *et al.* 1998).

In Poland, for example, the first information about the discoveries at Spiennes appeared in reports from the International Congresses of Anthropology and Prehistoric Archaeology in Brussels, written in 1873 by Count J. Zawisza, the father of Polish prehistoric research. After the discovery of the mine at Krzemionki Opatowskie, Spiennes became a reference point for the Polish site (Krukowski 1939: 49 and 122) and for various descriptions of European mining (Tabaczyński 1970: 275-276; 1972: 68; Hensel, Tabaczyński 1978: 140; Jażdżewski 1984: 165; Lech 1981: Fig. 2 and Table 2; 1991: 557-561; 1997: 611-623).

Consequently, Spiennes is now considered to be among the four or five most prominent sites of prehistoric mining in Europe, next to the mines at Grimes Graves, Cissbury, Rijckholt-St. Geertruid and Krzemionki Opatowskie (Lichardus *et al.* 1985: 37 and 441; Grooth 1991: 153; 1997;

Whittle 1996: 280; Lech, Leligdowicz 1998: 121 and 125; Lech, Longworth 2000: 19-20; Milisauskas, Kruk 2002: 214-215). As a result of many years of farming, as well as amateur digging and looting in the second half of the 19th and first decades of the 20th century, considerable damage has been done to the parts of the mine lying nearer the surface, so the Spiennes site is not as well preserved as the others just mentioned. However, shafts 1 and 2, excavated at the beginning of the 20th century by Baron A. de Loë, remain the deepest of all known European flint mine shafts and new excavations have shown the continued research potential of this complex site (e.g. Collet *et al.* 1997; Toussaint *et al.* 1997; Collet, Van Neer 2002; Defgnée, Collet 2003; Collet 2004).

5. THE TOPOGRAPHY OF THE MINING SITE

The extent of the exploited areas at Spiennes can be largely deduced from the mining waste present on the surface. Typical artefacts are flint nodules and their fragments, flakes and other flint waste, along with carbonated deposit, and chalk fragments. The latter material can only be taken into account in those parts of the site where the chalk is covered with a thick layer of silt. In some other parts of the site, this kind of evidence cannot be used, as the chalk naturally appears near the surface. In the northern part of *Camp-à-Cayaux*, prospecting for flint material on the surface was limited by the presence of meadows. The extent of flint extraction was also evidenced thanks to the numerous excavations and field observations made since the discovery of the site in the nineteenth century. It is worth noting that the investigations of the last decades have greatly increased our knowledge of flint extraction in Spiennes (Heim 2003; Fechner *et al.* 1993; Collet *et al.* 1997; Collet 2003a; Collet, Woodbury in press a, b and c). A systematic mapping of the scattered information as yet unpublished was recently undertaken and is still in progress. The documentation collected by F. Hubert and F. Gosselin and conserved at the Archaeological Service of the Walloon Region is helpful in this enterprise.

The present state of knowledge concerning the mining complex at Spiennes is shown on a map in figure 3. It must be clear that this is but a fragmentary view, merely reflecting the state of research in the different areas. However, this mapping generally confirms the physical extent of the mining field, on the basis of surface flint distribution. Shafts were noticed a long time ago in the area between the western and southern border of *Camp-à-Cayaux* and the *Centre de recherches archéologiques*, previously called the Museum (Fig.3.1 and 3; Loë, Munck 1891; Loë, Rahir 1929). Indications are less numerous beyond Harmignies Street but several exist (Fig. 3. 16-17 and Lefort, Cody 1953; Lefort 1954; Verheyleweghen 1962; Collet, Woodbury in press b). Moreover, in the northern part of *Camp-à-Cayaux* where meadows hide evidence of flint waste, investigations show a wider extension of flint extraction than was expected. For instance, about 20 shafts were discovered during the laying of a gas pipeline in 1992 (Fig. 3.15; Fechner *et al.* 1993). In the northwest sector 17 new shafts were identified during a rescue excavation in 2005 (Fig. 3.13). Flint exploitation is now confirmed as far as the first houses of the village along Harmignies Street, around the 57 m contour line. The presence of mines can be expected along the west border of *Camp-à-Cayaux* up to the 52 m contour line, where flint exploitation was noticed (Fig. 3.14). On the other hand, in other parts of the site, there is no reliable indication of flint extraction. Such is the case in the east and northeast part of the mining area of *Camp-à-Cayaux*, where only workshops were noticed by F. Hubert. In the same way, only the presence of workshops has been determined in the southeast part of *Camp-à-Cayaux*. In the same area however, some anomalies were reported during the digging of geological trial borings (Fig. 3.d). The geologists thought that they might have dug through flint shafts (Centre archéologique de recherche minière 1990: 4.17-4.19).

In *Petit-Spiennes*, north of the railway, flint extraction is confirmed by several excavations and some field observations (Fig. 3.6 and 3.12; Gosselin 1986; Heim 2003; Collet *et al.* 1997; Vandevelde, Hubert 1987; Collet 2003a). South of the railway, only some workshop areas have been found. In this case it seems highly probable that the absence of evidence is due only to the lack of investigation.

In the mining area called *Versant de la Wampe*, several mining features are known in the south western part of the *Petit-Spiennes* plateau, on the upper part of the slope surrounding the river La Wampe near the 70 m contour line (Fig. 3.9; Hubert 1976b). One isolated shaft was also identified near the top of the plateau on the same slope between the 72.5 and 75 m contour lines (Hubert 1971: 50-51).

6. MINING METHODS

Different mining methods were used at Spiennes, depending on the geological conditions of the deposits, as noted at other similar sites, such as Grimes Graves, Rijckholt-St.Geertruid and Krzemionki Opatowskie. The simpler exploitation units are poorly documented. Few are known and only one was the object of systematic excavations (Verheyleweghen 1953). Others were just explored by chance during the various campaigns (De Pauw, Van Overloop 1889-1890; Hubert 1976a; Collet, Woodbury in press c) or only partially excavated (Hubert 1969). Moreover the published data about the methods are only fragmentary (Verheyleweghen 1953).

On the western border of *Camp-à-Cayaux*, remains of extraction and production were excavated on the escarpment surmounting the river La Trouille. The remains showed horizontal niches dug straight into the slope. Each one was filled with a first layer composed of flint waste described as workshop, topped by a layer of yellow sediment mixed with blocks and chalk powder without any flint remains, either natural or knapped (Fig. 3.5 and 7). J. Verheyleweghen (1953) interpreted the system of exploitation as follows. The miners first dug short galleries about 2.5 m long and 0.7 m high into the slope. When this first extraction and production was exhausted, the roof of the first exploitation unit was pulled down to start a new one. A gallery dug directly into the slope was discovered in the vicinity during building work at the end of the 19th century (Fig. 3.2). Unfortunately L. De Pauw and E. Van Overloop (1889-1890) focused more on the chronological succession of the workshops connected with this gallery than on the exploitation method. This discovery may have suggested his model to J. Verheyleweghen. In 2005, a structure with a similar feature was observed in the northern part of the same escarpment (Fig. 3.14; Collet, Woodbury in press c). It is certain that this feature was dug to extract flint, as the chalk was left behind. The successive stages of the exploitation units are covered with layers of chalk in blocks separated by several layers of powdery chalk hardened by the circulation of the miners. The summit shows a collapse of natural chalk (Fig. 8). Unfortunately, dating was impossible, because no fragment of worked flint or other archaeological artefact appeared. Therefore the dating of these flint mines remains uncertain.

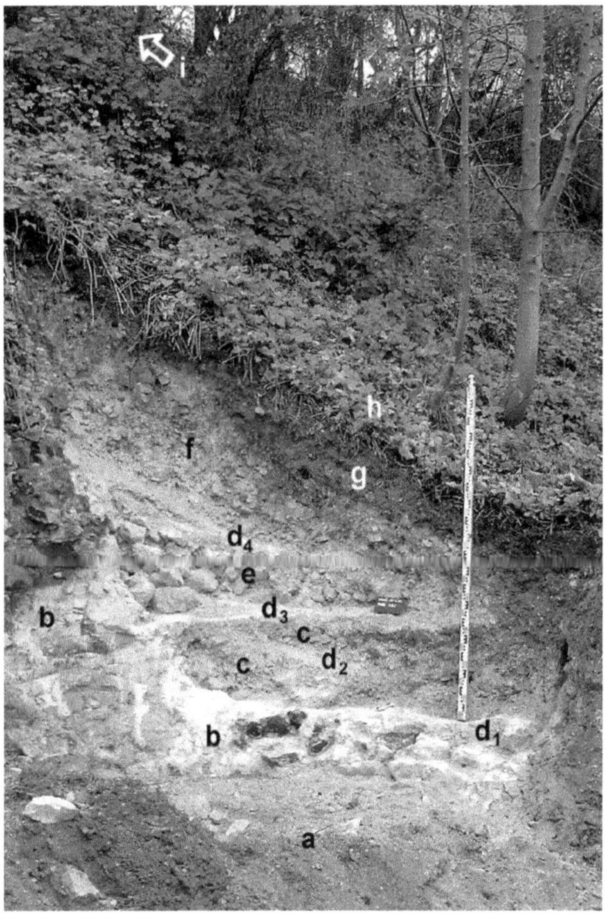

Fig. 8: Spiennes, Hainaut province. *Camp-à-Cayaux*. The slope of the La Trouille river valley. Flint exploitation remains found in 2005: a - modern embankment; b - natural chalk rock and flint seams; c - filling up deposit, chalk in powder; d - working floor of hardened chalk; e - filling up deposit, chalk in blocks; f - crumbled deposit of natural chalk; g - modern digging filled by humic sediment; h - slope; i - *Camp-à-Cayaux*. Photograph by M. Woodbury.

Open shafts mining pits are known near the rims of the two plateaus, where the silt layer is thin (Briart *et al.* 1868; Breuer 1930; Hubert 1969; 1976a). Some have been observed in the cross section of the railway trench, where the sediment covering the chalk is only 1 m thick (Fig. 3.a; Briart *et al.* 1868). Exploitation by trenches or large pits up to 3 m deep was also described at *Camp-à-Cayaux* (Fig. 3.4; Breuer 1930). In the 1965 excavation, where the chalk appears 1.2 m below the surface, two features were also considered as simple pits (Fig. 3.7). One was excavated down to 2.5 m, apparently without reaching the bottom of the shaft. In the 1975 excavation in *Petit-Spiennes*, three simple pits of 3.5 to 3.9 m depth were discovered (Fig. 3.10 and 9). They were dug in a place where the chalk is about 2.5 m deep. One of them has a preserved side exploitation (niche) 1.3 m long and high. The other

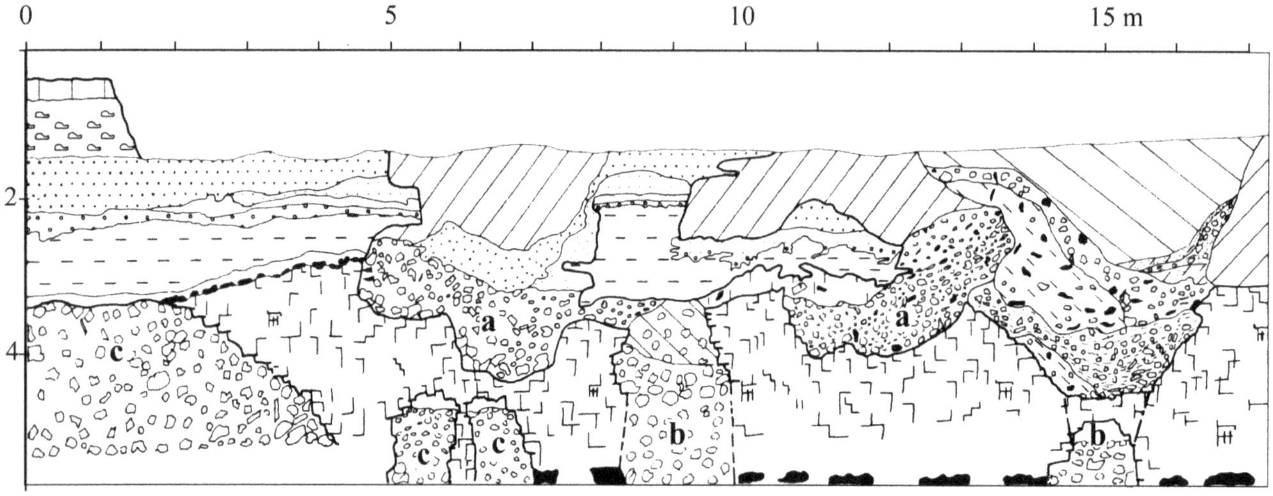

Fig. 9: Spiennes, Hainaut province. *Petit-Spiennes*. 1975 cross section near the railway trench. After F. Hubert (1976a): a - mining pits; b - shafts; c - galleries.

also shows such a lateral niche but the roof had collapsed and the original state could not be described.

Underground galleries are more fully documented. At Spiennes the shafts with underground galleries are between 3 to 16 m deep. Twelve mines were fully excavated and some more are currently being explored. For the rest, only the mouths of numerous access shafts are known. Underground mine galleries are located mainly in the centre of the plateau, where this is the only technique appropriate to exploit the flint seams. However, they also occur in areas where flint was exploited by other methods, like the digging of simple pits, as was recorded by some previous investigations (Fig. 9; Hubert 1969; 1976a), and as can be seen in the cross section of the railway trench (Fig. 5).

At *Petit-Spiennes*, two zones with underground galleries have been excavated: the first, since 1953 by the SRPH (Fig. 3. 6; Gosselin 1986), and the second in 1997-2004 (Fig. 3. 12; Collet, Van Neer 2002). Both areas reveal shafts from 8 m to 10 m deep depending on the thickness of the sediment. The miners had to dig down to 4.3 m, and 5.6 m respectively, before they could reach the geological layers of chalk, due to the natural sloping of the seams. A first flint seam usually appears around one meter below the chalk.

In these two areas, mines are known in which only the first flint seams struck upon during the digging were exploited. In shafts 53.2, 79.3 and 86.8.2 (area 1953), the first seam was occasionally extracted, while the second and the third were worked systematically. In shaft 11, explored between 1997 and 1999, only the first was worked. The ongoing investigation of shaft 20 also shows a shaft 10 m deep, with an exploitation of the first flint seams.

On the surface, shafts 80.4 and 53.2 are 1.7 m and 2.4 m wide in diameter. Deeper, the same shafts are slightly oval, measuring 0.7 m x 0.8 m in the case of shaft 80.4, and 1.1 m x 1.3 m across in the case of shaft 53.2. Shaft 11 is different, remaining wide right down to the exploitation level (Fig. 10). Its cross section is irregular with breadth varying from 3 m in surface to 2 m at 2 m below and then again 3 m wide at 3.8 m below and finally 2 m at 5.6 m below. Shaft 20 is even wider. At the surface it is oval shaped, with a diameter of 6.2 m x 5.2 m. At a depth of 3.4 m, it is 2.6 x 2.2 m wide then at 5.6 m depth it enlarges again to 3.7 x 3.1 m wide. In these two cases, the widening results from a deterioration of the shaft walls after the exploitation ended. In the case of shaft 20, it seems that the original diameter of the shaft was only a little more than 1 m across.

In the 1953 area, the shaft usually widens one meter below the top of the chalk level and forms a bell shaped room which can reach 3 m high (Fig. 11). From there, the galleries radiate outwards in several directions up to 4 m, or even to 5.5 m from the centre of the shaft. The height of the galleries varies from 1 m to 2 m. This is due to the fact

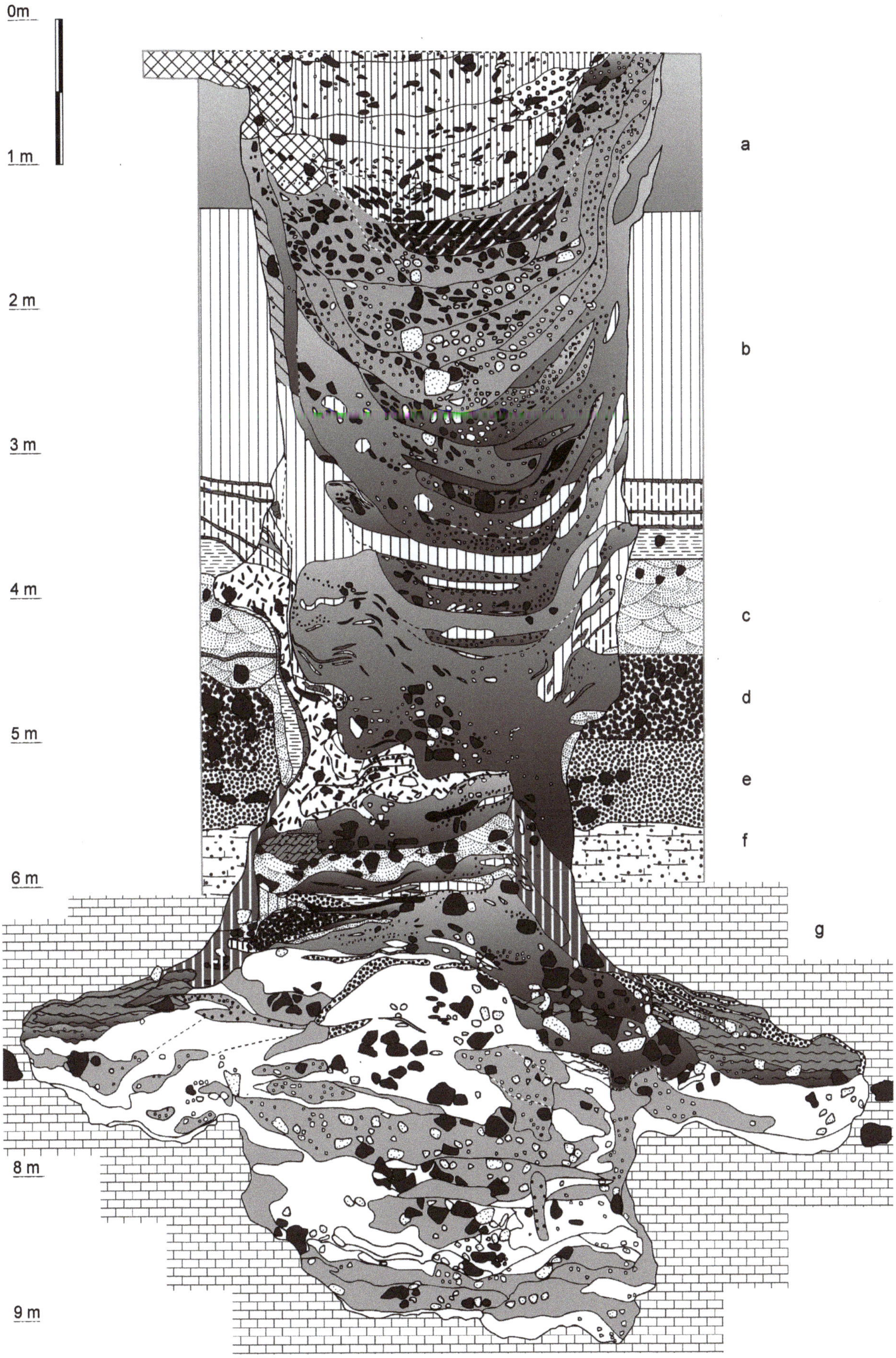

Fig. 10: Spiennes, Hainaut province. *Petit-SpienneS*. Shaft 11, excavated in 1997-1999. SW-NE cross section. a: clayey silt; b: loess; c: grey-blue sand; d: flint gravel, chalk, and sand; e: glauconiferous sand; f: chalk, and glauconiferous sand; g: chalk. Drawing by M. Woodbury.

Fig. 11: Spiennes, Hainaut province. *Petit-Spiennes*. Shaft 53.2. Bell-shaped room at the base of the shaft. Photograph by J.-L. Dubois.

Fig. 12: Spiennes, Hainaut province. *Petit-Spiennes*. Shaft 20. Flint picks cluster found in the exploitation level of the shaft. Photograph by Y. Gauvry.

that in these mines two different flint seams, separated by 1 m of chalk, were extracted from the walls at the same time. Excavations of shaft 11 in 1997-1999 (Fig. 3.12) showed short galleries (niches), at times only 0.7 m high (Fig. 10). Here only one flint seam was exploited. In the different units investigated, the surface worked covers 21 to 25 m^2. In both areas, the extracted flint consists of irregular nodules from 2.5 cm up to 34 cm long. They were mainly taken from the walls, and much more rarely from the bottom. The only mining tools found so far in the 1953 SRPH area and shaft 20 (area 1997, Fig. 12) are flint picks. The most common are bifacial, thin, with an almond shape. Much less numerous are the irregular and massive multifacial picks (see below). In the 1997 area, shaft 11 has yielded a mixed assortment of implements comprising a rich set of red deer antler tools and massive flint mining tools (Fig. 13).

In the area excavated by the SRPH since 1953, at least two different periods of mining were identified. The first dates from between 4400 and 3700 cal BC and the second from a period between 3500 to 2900 cal BC. Shaft 11, in the area excavated since 1997, is connected with the second period of exploitation. In shaft 53.2, dating from the first period, Michelsberg sherds from a bottle-shaped vessel were discovered in the shaft filling (Gosselin 1986).

In the centre of *Camp-à-Cayaux*, beneath the *Centre de recherches archéologiques*, there are shafts 15 m to 16 m deep (Fig. 3.3, 14 and 15). The explored underground area covers about 250 m^2. Like elsewhere at Spiennes, the two access mouths of the excavated shafts 1 and 2 are about 1 m wide. The miners had to dig 5 m to reach the chalk but here they did not exploit the first flint seams and passed through 15 flint layers before reaching the flint they were seeking (Fig. 15; Hubert 1978: 8, 12). At that depth, a twin flint seam occurs, consisting of large slabs 1 m wide and up to 1-2 m long and about 20 cm thick. Only the lower level was extracted, while the other was left *in situ*, apparently in order to protect the galleries. As at *Petit-Spiennes*, the bottom of the shaft forms a room surrounded by chalk pillars. The mine workings radiate outwards in several directions for a distance of up to 4 m. These measurements are estimations, as it is sometimes difficult to know which shaft is connected with the galleries. Some may have even been

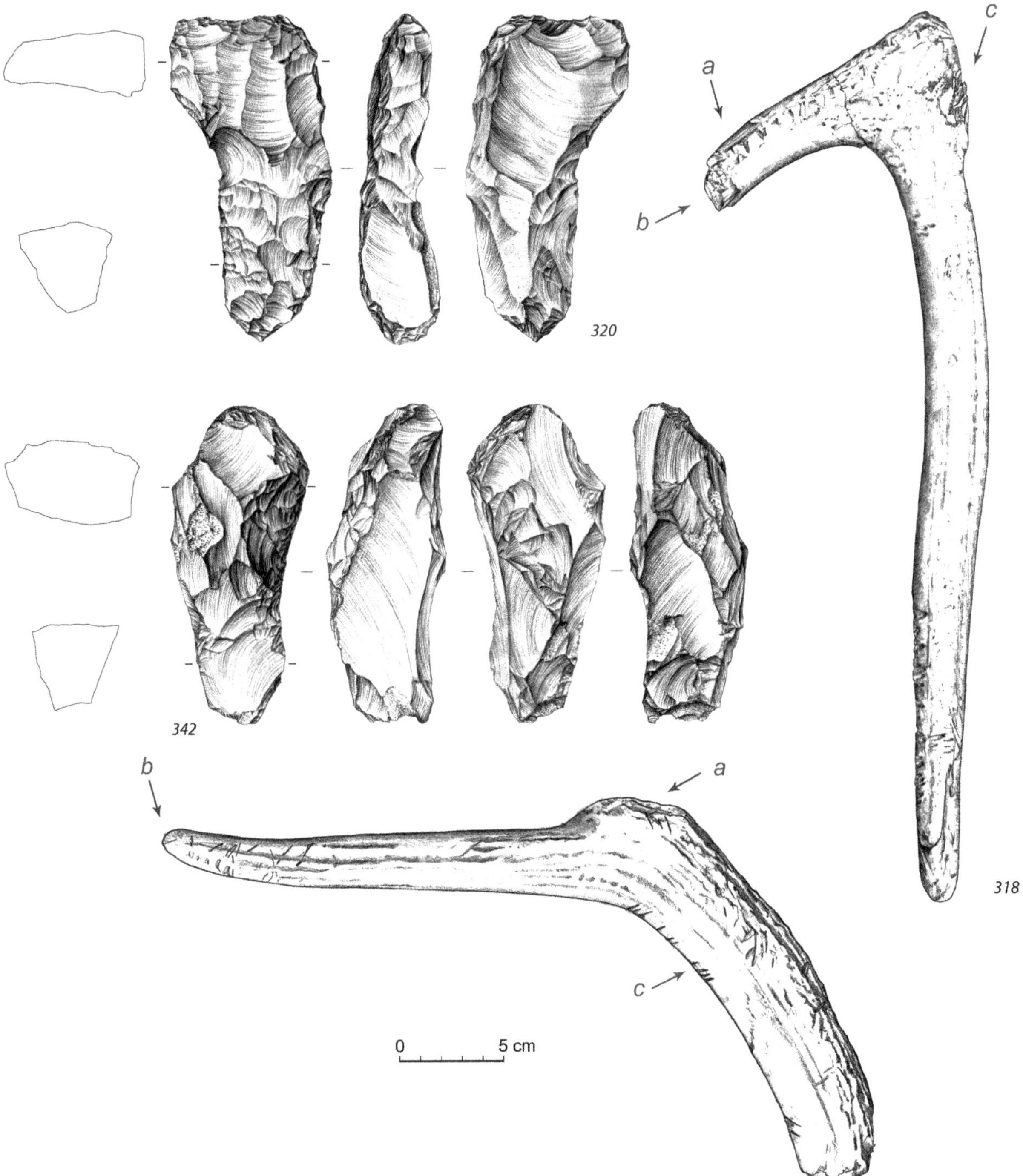

Fig. 13: Spiennes, Hainaut province. *Petit-Spiennes*. Shaft 11. Mining tools: 309 - a-b: wear marks, c - undefined parallel notches; 318 a, b, c - wear marks; 320 and 342: drawings by E. Gumińska; 309 and 318: drawings by P.-Ph. Sartieaux.

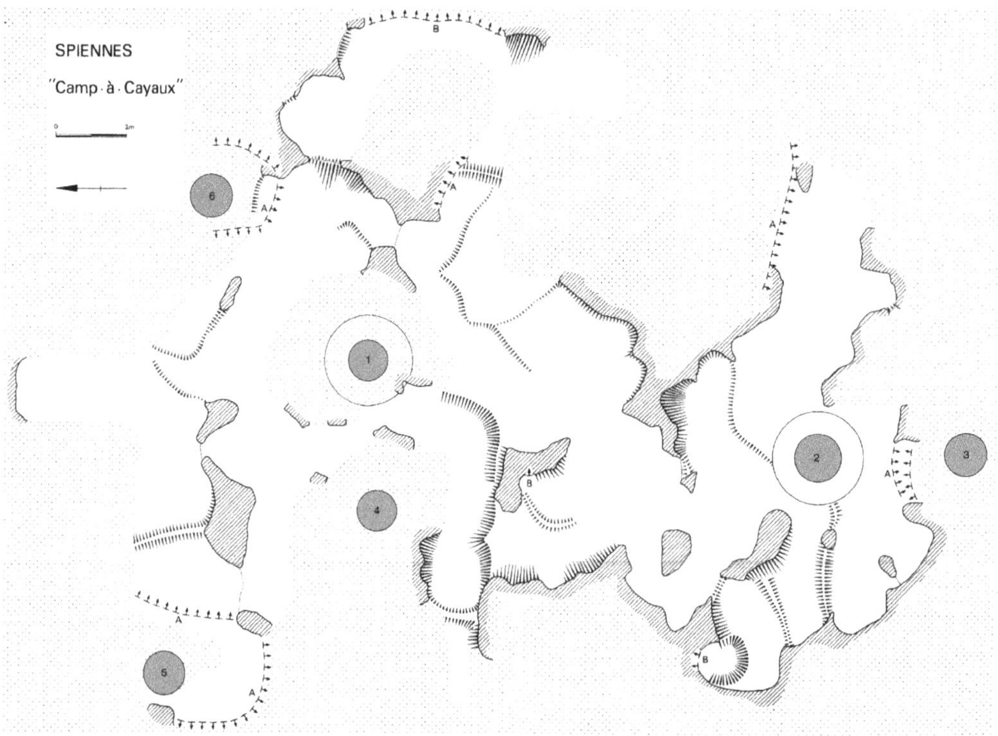

Fig. 14: Spiennes, Hainaut province. *Camp-à-Cayaux*. Shafts 1 and 2. Plan of the shafts base with galleries. After F. Hubert (1978).

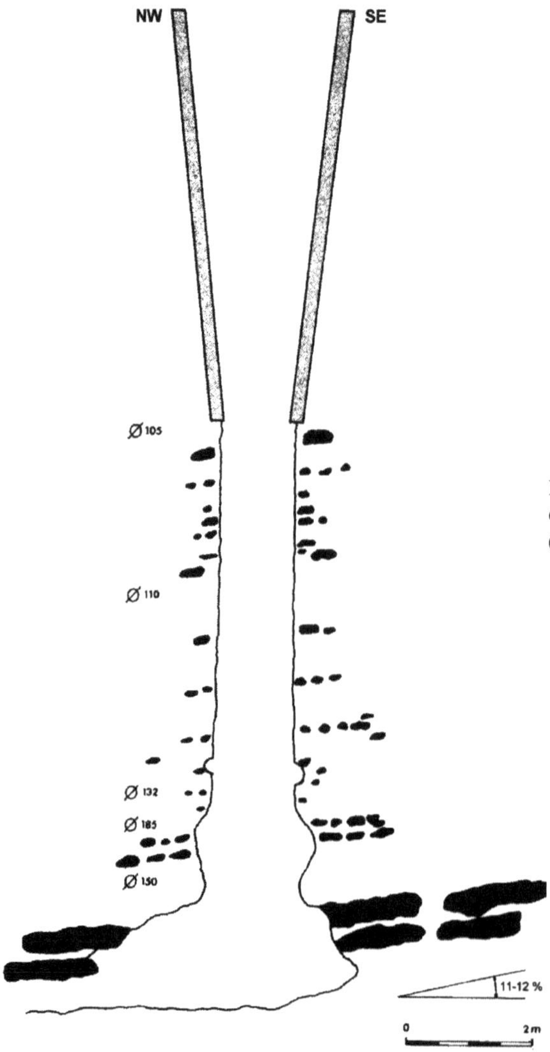

Fig. 15: Spiennes, Hainaut province. *Camp-à-Cayaux*. Shaft 2. Cross section. After F. Hubert (1978).

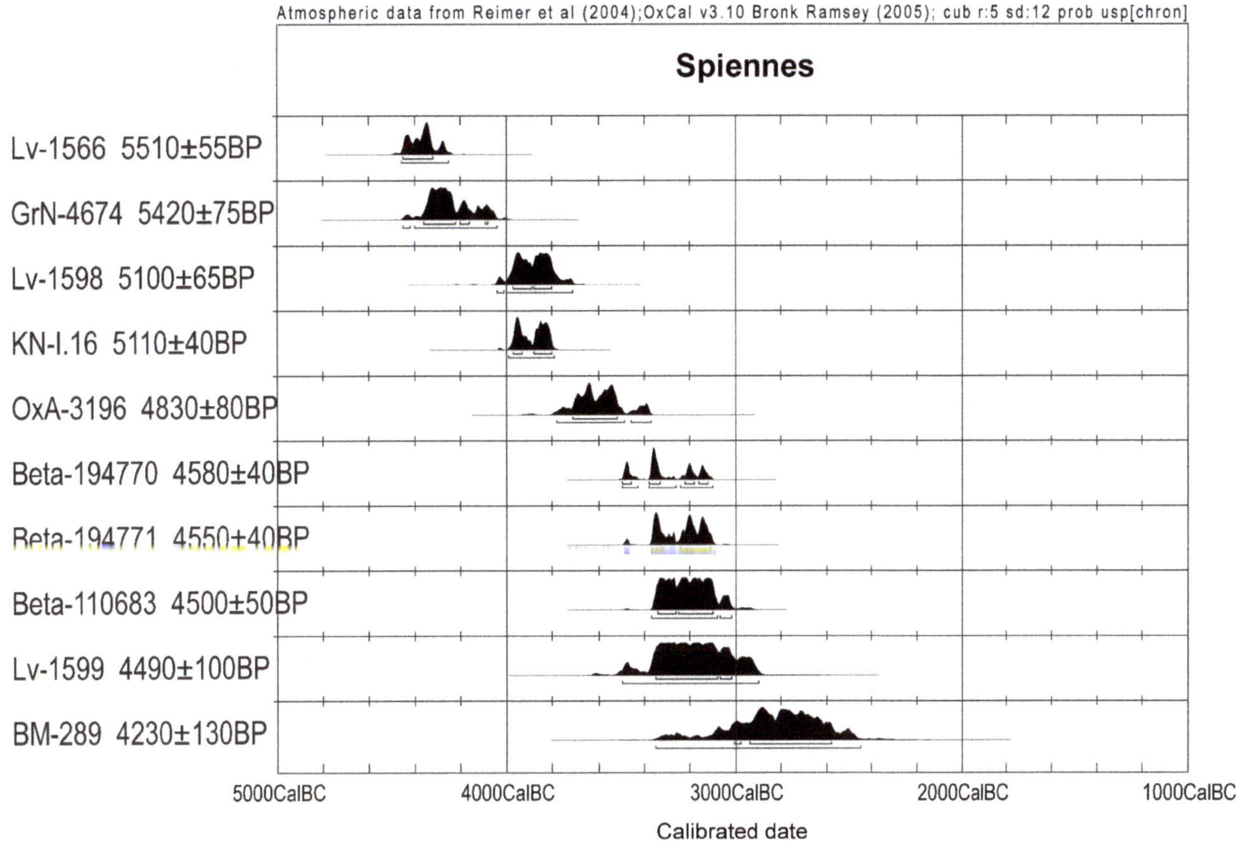

Fig. 16: Spiennes, Hainaut province. The prehistoric mining complex. Radiocarbon dates.

8-9 m long. The height of the levels exploited varies from 0.7 m to 1 m. Flint was extracted from the roof by digging the chalk under the twin slabs. The area exploited in each of the two, probably fully investigated, mines can be estimated at around 45 m². The mining tools used to dig these galleries comprise only flint picks of various shapes. No radiocarbon date is available for these mines. The only element of dating is given by fragments of a Michelsberg ceramic vessel found near the bottom of an unexplored shaft located near shaft 1 (Hubert 1978: 38). It is generally accepted that these mines were exploited by the Michelsberg communities between 4300 and 3700 cal BC.

In the same area, only 20 m from the limits of the underground galleries belonging to shafts 1 and 2 described above, five shallow shafts were explored (Fig. 3.11; excavation 1990: shafts 2, 4, 5, 6 and 7) over an area of 100 m². From centre to centre, the distance between these shafts was 3 to 5 m. Their depth was only 3 to 3.5 m (Bonenfant, Auzou 1993; Bonenfant 1993). They were 1 to 2 m wide at the mouth and 1 m or less at a depth of 1 m. The flint was exploited from niches 1 m long and 1.2 m high. Flint picks were the basic mining tool, though fragments of red deer antler were also found (Bonenfant, Auzou 1993). No radiocarbon date is available but from the discovery in shaft 7, at a depth of 1 to 1.3 m, of a Michelsberg bottle-shaped vessel broken *in situ*, we may conclude that these mines also date from the Michelsberg period.

So far, the underground exploitation of the mining area of the *Versant de la Wampe* has only been partly excavated (Fig. 3.9). Only some of the mine workings were explored during the excavation of three trenches which had been dug to establish the layout of the ditches of the Michelsberg enclosure. The excavation of the ditches went down to 2 m. This was enough to identify the remains of mining features: two shafts little more than 1 m wide and two galleries with roofs located only 1.3 to 1.7 m under the surface.

7. DATING

The few available radiocarbon dates indicate that the mining field was exploited between 5510 to 4230 BP, then from 4400/4200 to 3350/2900 cal BC (Fig. 16). We don't know if exploitation was continuous or not. The mining activities were thus contemporaneous with the Middle Neolithic period in Belgium but continued long after. The pottery discovered in the mining areas shows that the site was mainly in use during the Michelsberg period (4300-3700 cal BC). The context of the find shows that this pottery was, at times, strictly connected with mining exploitation or at least contemporaneous with it (Colman 1957; Hubert 1978; Gosselin 1986; Hubert, Soumoy 1993). Some Michelsberg potsherds were recently discovered in the close context of the chipping floor excavated at Spiennes - Lot 51c. The remains of earlier periods are very scarce and not connected with mines. Those of later periods are also poorly documented. Most potsherds from these later occupations are not assigned to specific culture and can date from the Late Neolithic to the Bronze Age. Some Late Bronze Age and Iron Age finds were clearly identified at the mining sites at *Camp-à-Cayaux* and *Petit-Spiennes* (Mariën 1961). It remains unclear whether a direct or indirect relationship with mining exploitation can be established. Nevertheless, according to F. Hubert (1976c), one of the Late Bronze Age features is connected with flint processing. So far however, incontrovertible evidence is lacking for mining later than the end of the 4th millennium cal BC.

8. STUDY OF FLINT WORKING AT THE SPIENNES MINING FIELD

Not much is known about flint working in the Neolithic Spiennes, as compared to such prehistoric flint mining sites as Grimes Graves in England, several of the Polish mines or Jablines in France. Although the first articles devoted, to some extent, to flint working at Spiennes were written as early as the 19th century, in connection with the research done by L. De Pauw and E. Van Overloop (1889-1890), archaeologists of later periods showed little interest in the flint workshops before the middle of the 1960s.

F. Hubert was the first archaeologist to display more interest in the flint finds at Spiennes. In 1965, he excavated four workshops, but published only their general characteristics, paying greater attention only to the typology of the basic categories of artefacts: cores, blades, axes, chisels, picks and *tranchets* (Hubert 1969). Compared to earlier works, F. Hubert's article (1971) on the excavations of a Michelsberg enclosure on the *Petit-Spiennes* upland devoted considerable attention to flint material. But it was only from 1981 on, with the excavations at *Petit-Spiennes* carried out by SRPH, that all the waste flint material obtained during the excavations was collected. A significant part of the material from *Petit-Spiennes* was then analysed by F. Gosselin (1986). The published results of her studies, constitute, so far, the best presentation of flint material from the mining complex at Spiennes, though they include only material obtained during excavations of shaft fillings 53.2 and 80.4 and of underground extraction areas of these and several neighbouring shafts at *Petit-Spiennes*.

In 1997, H. Collet began excavations at Spiennes, paying particular attention to the flint material. From 1997 to 2004 she concentrated on the *Petit-Spiennes* shafts, but in 2005, while doing some short-term rescue work, she excavated large areas of flint workshops at the north edge of the *Camp-à-Cayaux* mining field (Collet, Woodbury in press a).

8.1. Flint finds from Spiennes

The results of the Spiennes mining complex excavations completed so far make certain generalisations possible regarding the basic trends of flint production and the categories of artefacts. However, little can be said about the flint workshops and the techniques or technology of flint working, which left behind a great mass of flakes and flint waste. We have at our disposal only some very modest data on the flint inventories, obtained from F. Hubert's (1969; 1971) excavations, and the recent SRPH research at *Petit-Spiennes*.

The flint mined at Spiennes served to make blade blanks from cores, and a small number of flake blanks. The production of axe blades is especially well substantiated and, to a much lesser degree, that of other core tools, of which chisels – *ciseau* – though very rare, are a typologically distinct category (see Brézillon 1971: 195, 196). Mining tools form a separate group. Picks predominate, while other mining tools consist of various failed specimens obtained in the process of producing various implements, such as axe blades, and then adapted for the job of flint extraction as the need arose (Lech 1982-1983: 24-26; 1991: 560-569).

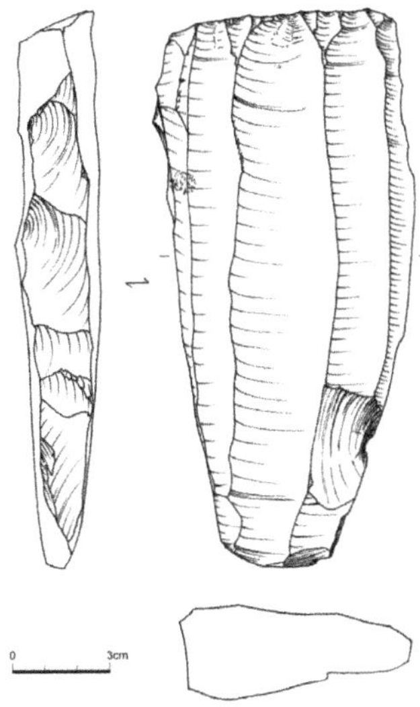

Fig. 17: Spiennes, Hainaut province. *Camp-à-Cayaux*. Flint core for blades from workshop 7. RBINS: Stevens collection. Drawing by A. Hauzeur.

Flake cores seem to be a rarer category at Spiennes than blade cores. The specimens are varied and difficult to attribute to standardised categories. Excavations so far have turned up flake cores more often at *Petit-Spiennes* than at *Camp-à-Cayaux* (Gosselin 1986: 90-91, 114-115).

Blades and blade fragments occurred frequently at *Camp-à-Cayaux* and much more rarely at *Petit-Spiennes*. The longest regularly shaped blades were obtained from cores prepared from flint extracted from the deep shafts of *Camp-à-Cayaux*. Both regular and failed specimens occur.

Cores and blades

Single platform blade cores are among the most important categories of finds from the mining complex at Spiennes (Fig. 17). Many of them have standardised features and could be associated with those workshops which exploited flint from the deepest shafts. The largest regular standardised cores would have been prepared from fragments of flint blocks extracted from the deep shafts at *Camp-à-Cayaux*. They are similar to the famous '*livres de beurre*' from Grand-Pressigny. Compared to the classic '*livres de beurre*', the regular cores from Spiennes are smaller and less carefully prepared; also the variety and range of flint knapping techniques was more limited. It seems that more blades were obtained in the process. After extraction some of the cores were immediately converted into axes. Core knapping was preceded by preparation of the pre-core. During this process, much attention was paid to the preparation of the blade's striking surface. The trimming technique was generally used. The core exploitation methods used at Spiennes were the same as at Grand-Pressigny, and at workshops 1 and 3/1960 in Sąspów (Dzieduszycka-Machnikowa, Lech 1976). Blades were struck off from the partly rounded striking surface until it became flat, at which point it had to be partly rounded again.

At the Spiennes mine complex there were areas where blade cores were prepared and worked, for instance near shafts 1 and 2; and also in the southwest part of *Camp-à-Cayaux*, and at *Petit-Spiennes*, near shafts 20, 28 and 53.2. We also know of areas where no cores were found, such as the north-northwest part of *Camp-à-Cayaux*, where rescue excavations were carried out in 2005.

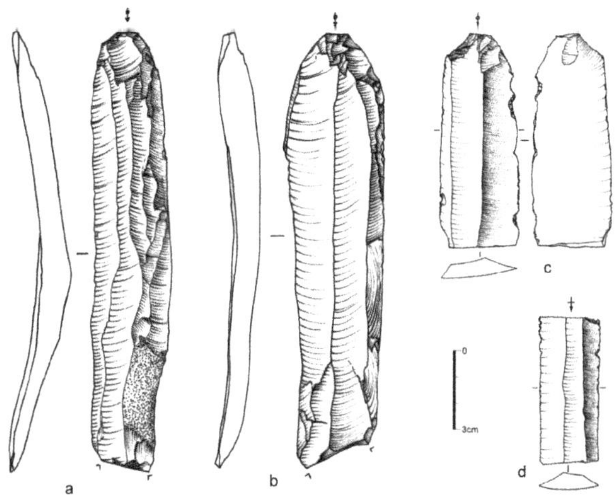

Fig. 18: Spiennes, Hainaut province. *Camp-à-Cayaux*. Blades. RBINS: Stevens collection. Fig 18 a-b: drawings by A. Hauzeur; c-d: drawings by A.-M. Wittek.

Measuring a sample of blade scars on such cores from Ch. Stevens' collection at the RBINS (Fig. 18), it was determined that most of the blades were between 15-21 cm in length, 25-35 mm wide and 7-10 mm thick. The flint material at *Camp-à-Cayaux* was also characterised by large core trimming blades and further blades with traces of core preparation (see also Vanmontfort *et al* in press).

In the *Petit-Spiennes* area, regular blades like the ones from *Camp-à-Cayaux* are rare. Drawings of cores from F. Gosselin's publication (1986: 92, Fig. 31) confirm the production of blade blanks of up to 20 cm in length.

Axe blades and chisels

Among specimens that underwent preliminary working, early roughouts, and roughouts axe blades were the most common. All axe blades were bifacial. An analysis of the

Fig. 19: Spiennes, Hainaut province. *Camp-à-Cayaux*. Two axe blades from the area of the Centre de recherches archéologiques. Photograph by M. Woodbury.

roughouts' earliest steps and of the roughouts themselves indicates large variations in the group, both in morphology and size. The largest known roughout of an axe blade from Spiennes comes from *Camp-à-Cayaux*; it was published by A. de Loë (1914). This unusual specimen is 28 cm long and was carefully prepared to the stage which precedes polishing. It was found in or just next to a flint knapping workshop excavated in 1913, located near shafts 1 and 2. Two finished roughouts of axe blades of about 20 cm (Fig. 19) were also recovered in the same area in 2006. Both specimens were made of massive flakes struck off from large flint nodules which had been extracted from deep shafts in this area of the mining field.

During excavations of the southwest part of *Camp-à-Cayaux* carried out in 1965, F. Hubert (1969) found 62 early roughouts and roughouts of axe blades and fragments (he also found 118 blade cores and 17 flake cores). The biggest is nearly 22 cm long. Large axe blades were also produced at *Petit-Spiennes* (Fig. 20). Because of the difference in the character of the raw material exploited here, they were smaller than the largest specimens at *Camp-à-Cayaux*.

Among the flint tools from Spiennes we also distinguish chisels or *ciseaux*. Generally speaking, these are tools prepared using the same technique as with axes but slimmer in shape (Brézillon 1971: 195 and 196). Chisels were probably among the tools used during activities carried out within the mining field area, and altogether they comprise only a small part of the artefacts produced at the Spiennes workshops. Another very rare tool at Spiennes is the *tranchet*, a more typical tool in Middle Neolithic domestic context.

Mining tools: picks and others

The most important mining tool at Spiennes was the pick, used primarily for digging chalk. Excavations have yield-

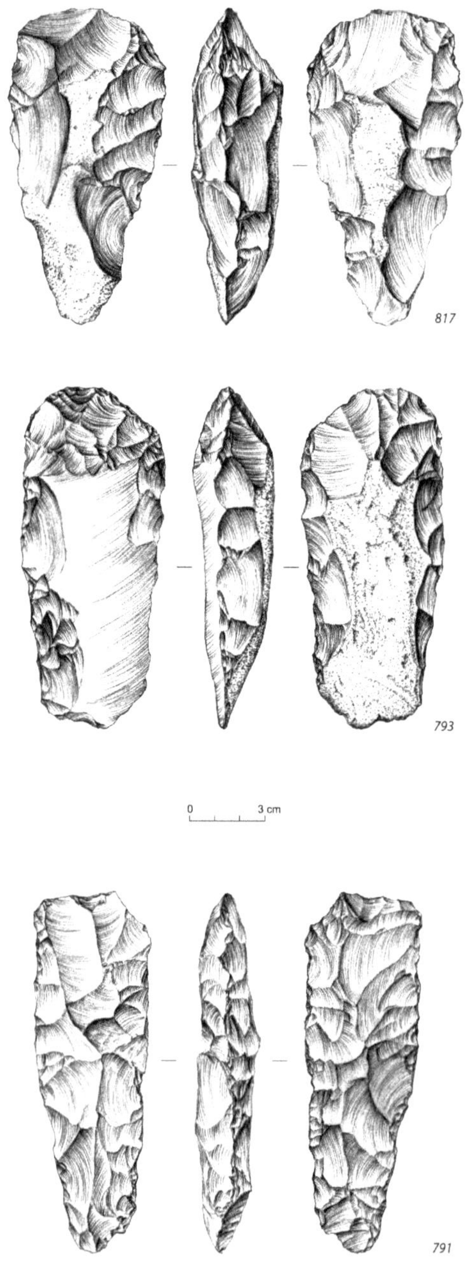

Fig. 20: Spiennes, Hainaut province. *Petit-Spiennes*. Shaft 11. Axe blade roughouts. Drawings by E. Gumińska.

ed many tools belonging to this category; most of them found in shafts 1 and 2 at *Camp-à-Cayaux* during excavations carried out by Baron A. de Loë. In his first campaign, about one thousand picks were found (Fig. 21) and in the second another 500 (Loë 1913: 45; 1914: 35).

F. Hubert (1978: 24-31) analysed 117 picks found during the excavation of underground galleries between shafts 1 and 2 conducted by M. Lefort in 1948-1954. He divided them into 7 different types. The most common were fusiform picks (*Pics fuselés typiques*) – 45 specimens, that is 48.9%. F. Gosselin (1986: 54-87) examined 289 picks from three shafts in the *Petit-Spiennes* area: 53.2, 79.3 and 80.4. She divided them into two main categories: the multifacial (*multifacial)* including 48 specimens, and the bifacial (*bifacial*) comprising 241 artefacts. The latter is the most common in this area. In shaft 20, whose exploitation level was partially excavated, about one hundred almost-whole picks, 35 large fragments, and several hundreds small chips were distinguished. Almost all of them, following F. Gosselin's typology, were classified as bifacial. Some of them were only 'functional' picks, i.e. failed early roughouts and roughouts of axe blades used in mining work.

The large numbers of picks recovered from the mine pits and shaft fillings show how important this tool was for the flint miners at Spiennes. These numbers also suggest that picks wore off relatively quickly and had to be frequently replaced by new ones. For example, a broken pick and fragments were found at the bottom of the backfill in shaft 20. The refitting of these artefacts gives a good idea of how tools were progressively worn away during work. It shows that the fragments came off bit by bit, and not all at once in one strike (Fig. 22).

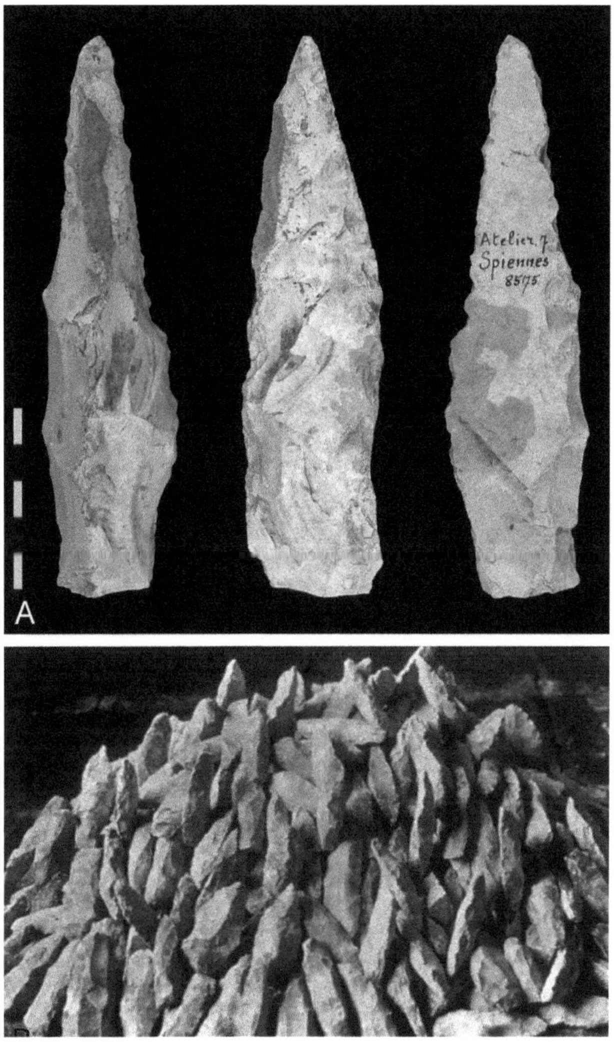

Fig. 21: Spiennes, Hainaut province. *Camp-à-Cayaux*. A: Flint pick from workshop 7. RBINS: Stevens collection. B: Flint picks from 1912-1914 excavation of Shafts 1 and 2. A: Photo E. Dewamme, RBINS; B: After A. de Loë (1925).

Fig. 22: Spiennes, Hainaut province. *Petit-Spiennes*. Shaft 20. Refitted bifacial flint pick. Photograph by M. Woodbury.

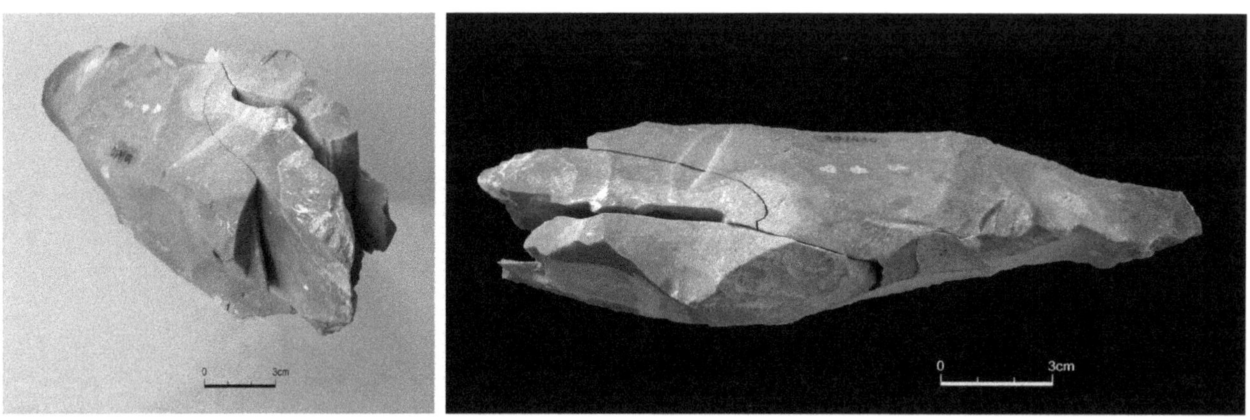

Other flint tools were also used in the process of mining. Usually these were previously-discarded roughouts of axe blades. They were adapted *ad hoc* for the job of mining in chalk. Evidence of transforming exhausted blade cores into *ad hoc* 'functional' picks must also be noticed.

8.2. Flint knapping workshops

Among the tools used for flint knapping at Spiennes were sandstone and quartzite hammers, though flint hammers are equally numerous. This group of tools is extremely variable in size and weight, with two specimens exceeding 0.5 kg and many considerably lighter ones.

Another characteristic category of finds connected with the final phase of making axe blades and chisels are polishing stones or *polissoirs*. 49 such objects have been reported, not counting the specimens noted in the Michelsberg enclosure excavations and in the recent excavations (Collet 2000). Most come from the *Camp-à-Cayaux* mining field. *Polissoirs* are present - mostly in fragments - but we don't know about their proportional representation. The largest comes from *Camp-à-Cayaux* and was found in the 19th century during some unidentified 'digging'. It measures 41 cm long, 24.3 cm wide, 29.9 cm thick and offers four surfaces for axe polishing (Delvaux 1885-1886: 197).

Another tool connected with flint workshops are retouchers. They are a rare find and it is not always possible to determine whether specific traces of utilisation may not be the result of the incidental use, as a makeshift hammer. Nevertheless, their occurrence at Spiennes is certain. Another unusual category of finds connected with flint knapping are bone punchers, used for indirect percussion to remove blades from cores. They are known from older excavations, and F. Hubert (1969: 28) also found a deer antler puncher when excavating in 1965 workshop II. F. Hubert's excavations in 1965 represent the only extended studies of flint workshops in the history of the Spiennes excavations. He completely excavated workshops I and III, as well as parts of workshops II and IV, and published the general results of his work (Hubert 1969). In workshops I, II, and III he found sherds of Michelsberg culture pottery. Only flint tools and objects made from other raw materials were published from these excavations. Large samples of archaeological material from several other workshops were obtained recently by H. Collet.

8.3. Initial analysis of flake samples from Lot 51C, feature 5

Towards the end of January and during February, 2005, rescue excavations were carried out on a building lot lying in the north-northwest part of *Camp-à-Cayaux*. The digging of the 150 m^2 trench lasted three weeks, during which time nearly 50 percent only of the features and materials were explored (Collet, Woodbury in press a). This revealed the presence of *in situ* workshop places. Under emergency conditions, all the flint from the excavated areas was removed '*en bloc*', packed with the sediment to be sieved later. After it had been collected, 21 features were located, of which 17 were definitely shafts (Fig. 23).

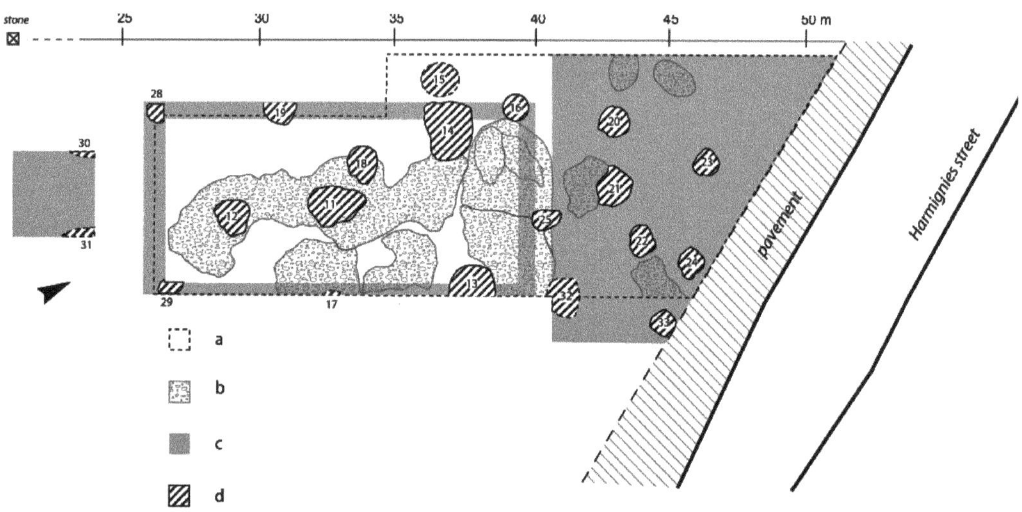

Fig. 23: Spiennes, Hainaut province. *Camp-à-Cayaux*. Plot 51c. Plan of the 2005 excavation: a - excavated area; b - distribution of flint material after removal of topsoil and cleaning; c - house area and parking area in front of house; d - shafts. Drawing by M. Woodbury and A. Hauzeur.

Fig. 24: Spiennes, Hainaut province. *Camp-à-Cayaux*. Plot 51c. Sample of the material explored during the 2005 excavation. Differentiation of size. A - flakes; B - sieved chips. Photograph by M. Woodbury.

An initial study of 1233 flakes and flake fragments was carried out with preserved butts > 20 mm, together weighing 4985 g (Fig. 24). The sample was organized into dimensional classes, taking into consideration the maximum size of flakes (Fig. 25), following the divisions adopted in the classification of workshop material from Sąspów and Grimes Graves (Lech, Longworth 2000; 2006). Independently, the specimens were also divided according to thickness (Fig. 26), as proposed by A. Augereau (1995). This is the first, preliminary comparison of the results of two classifications carried out on the same, small sample. A comparison of diagrams from Fig. 25 and 26 shows much greater variations in the frequency of specimens in the three size categories of large, ≥ 80 mm, medium, 50-80 mm, and small 20-50 mm flakes than in the three thickness categories, i.e. very thick, > 15 mm, thick, 5-15 mm and thin, < 5 mm flakes. As mentioned earlier, from Lot 51c we only know of workshops preparing flint axe blades and there are no indications that blade cores were prepared there. Flake cores, if they occurred, seem rare.

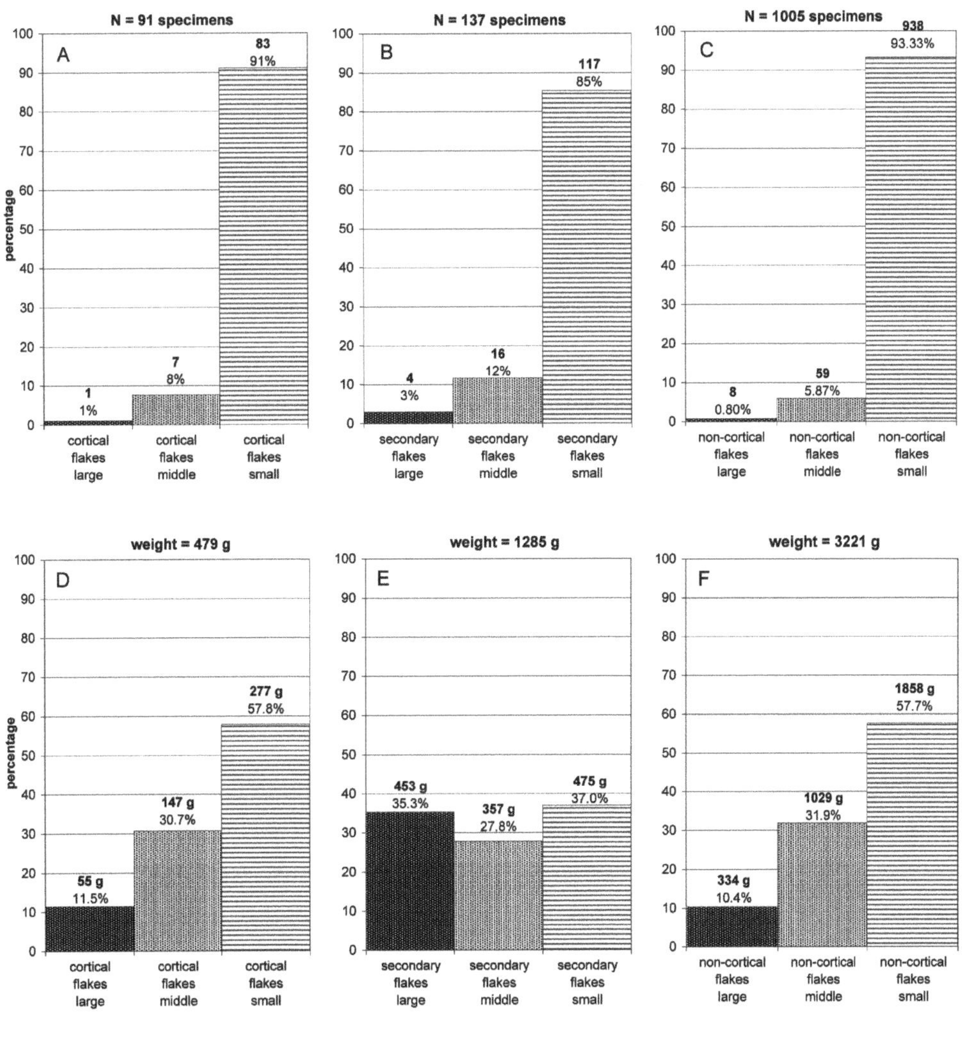

Fig. 25: Spiennes, Hainaut province. *Camp-à-Cayaux*. Plot 51c. A sample of flint material from workshop 5/2005. Differentiation of basic attributes of flakes > 20 mm on basis of the size (large > 8 cm, middle 5-8 cm, small 2-5 cm) and upper surface of specimens: A – total cortical flakes; B – secondary flakes (partially cortical); C – non-cortical flakes.

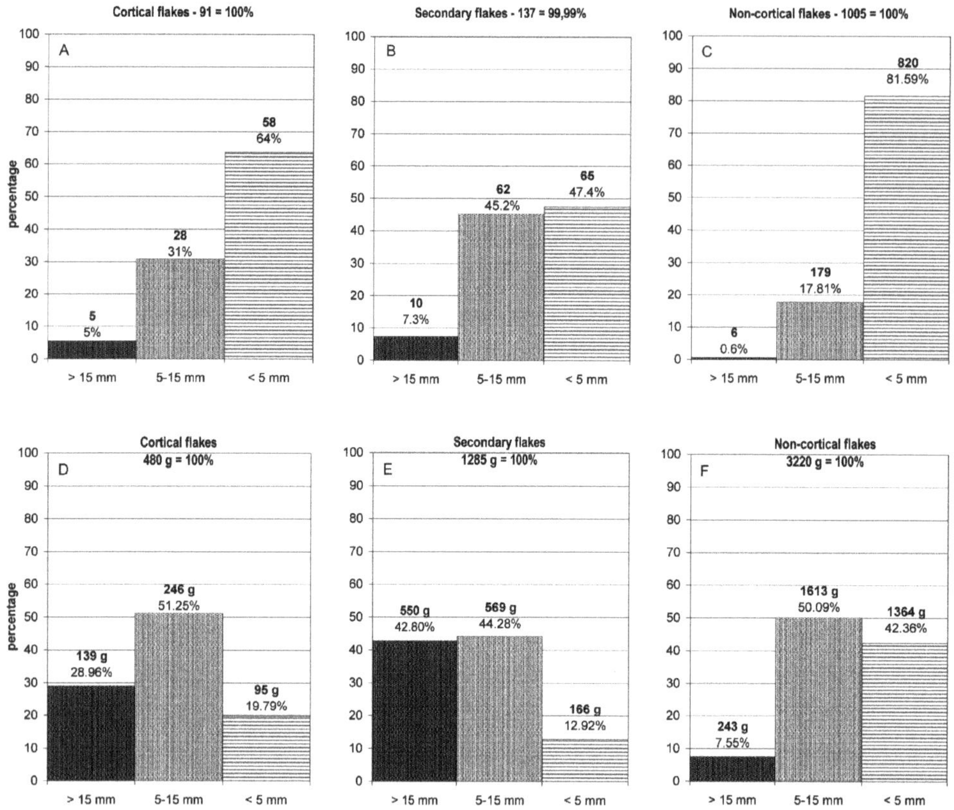

Fig. 26: Spiennes, Hainaut province. *Camp-à-Cayaux*. Plot 51c. A sample of flint material from workshop 5/2005. Differentiation of basic attributes of flakes > 20 mm on the basis of the thickness and upper surface of specimens. (very thick > 15mm, thick 5-15mm, thin < 5mm).

	Large flakes ≥ 80 mm	Middle flakes 50-80 mm	Small flakes 20-50 mm
Cortical flakes	55 g ?	20.6 g ?	**3.3 g**
Secondary flakes	113.2 g ?	22.3 g ?	**4.1 g**
Non-cortical flakes	41.7 g ?	**17.4 g**	**2 g**

Fig. 27: Spiennes, Hainaut province. *Camp-à-Cayaux*. Plot 51c. A sample of flint material from workshop 5/2005. Average weight of basic categories of flakes > 20 mm on the basis of the size and upper surface of specimens.

	Very thick flakes > 15 mm	Thick flakes 5-15 mm	Thin flakes < 5 mm
Cortical flakes	28.7 g ?	8.8 g ?	**1.6 g**
Secondary flakes	13.7 g ?	**9.2 g**	**2.5 g**
Non-cortical flakes	40.5 g ?	**9 g**	**1.7 g**

Fig. 28: Spiennes, Hainaut province. *Camp-à-Cayaux*. Plot 51c. A sample of flint material from workshop 5/2005. Average weight of basic categories of flakes > 20 mm on the basis of the thickness and upper surface of specimens.

The sample is notable for the predominance of small flakes of 2 to 5 cm from each category (N=1138), totalling 92.3% of all flakes (Fig. 25). However, with regard to weight their share is much more modest – 52.4%. At the same time, though not as evidently, there is a predominance of thin flakes when it comes to numbers - 76.5% of the total -, though they constitute only 32.6% in regard to weight. It is the opposite case with large flakes, only 13 (1%) in number but they constitute 17% in terms of weight, and very thick flakes, altogether 21 (1.7%) in number but they constitute 18.7% in terms of weight (Fig. 26). The predominance of the whole non-cortical flakes (N=1005) when compared with the total of cortical flakes (N=91) is only slightly smaller in number - 81.5% for 7.4% - and still clear, though lesser in the case of weight – 64.6% for 9.6%. Generally, the number and weight analyses confirm the relations determined earlier for the Grimes Graves mine. The analysed sample was too small to warrant an evaluation and comparison with the other weight averages of large, and medium sized flakes, and the very thick flakes. Moreover, studies to determine how both these features are correlated (Fig. 25-28) have still to be conducted. However, it should be noted that the average weight of a specimen from the group of partially cortical flakes (9.4 g) is nearly twice the average weight of a totally cortical flake (5.3 g) and even larger

Fig. 29: Spiennes, Hainaut province. *Petit-Spiennes*. Shaft 20. A sample of flint nodules abandoned by miners in the exploitation level. Note traces of removed flakes on some nodules. Photograph by J. Lech.

than the average weight of a non-cortical flake (3.4 g). Given these preliminary considerations, it would seem that the first flakes detached from the blank (nodules or large flakes) were intended to peel off the surface rather than to check its quality, and to prepare it for the basic procedures of the following shaping process. Therefore, these flakes were neither the largest nor the heaviest.

When the nodules, or their large fragments were being shaped, the flakes removed were larger than in the first phase. These are secondary flakes, with partially cortical surfaces. During the last phase, when the early roughouts, and the roughouts of axe blades, were being formed, many small flakes and microflakes (with a maximum size of 10 to 15 mm) were produced; the last two categories were produced in large numbers during the final shaping of the edges of the axe roughout.

The sample examined here remains very small but seems to support the argument just presented, which is based mainly on an analysis of material from workshop complexes at the Grimes Graves mine (Lech, Longworth 2000; 2006).

8.4. Organisation of the flint working process

Studies of the mine complex at Spiennes and of flint working in communities of the Michelsberg culture in the Spiennes region and neighbouring areas, allow us to draw only a preliminary sketch and some general conclusions about the organisation of flint working here, conclusions much more modest than those deriving from the Grimes Graves mine. At Spiennes, flint nodules were extracted from various levels and differed in size and quality. In the case of the recently excavated shafts at *Petit-Spiennes*, which provide fresh observations, it was determined that the extracted nodules were cleared roughly of chalk, checked for quality by removing the largest protuberances and simple flakes (Fig. 29). The good nodules were transported to the surface; the rejected ones were left underground. Among the latter were all the small nodules (Fig. 30). The analysed sample of flint material comes from the exploration of a 15m^3 waste dump left at the bottom of shaft 20 at

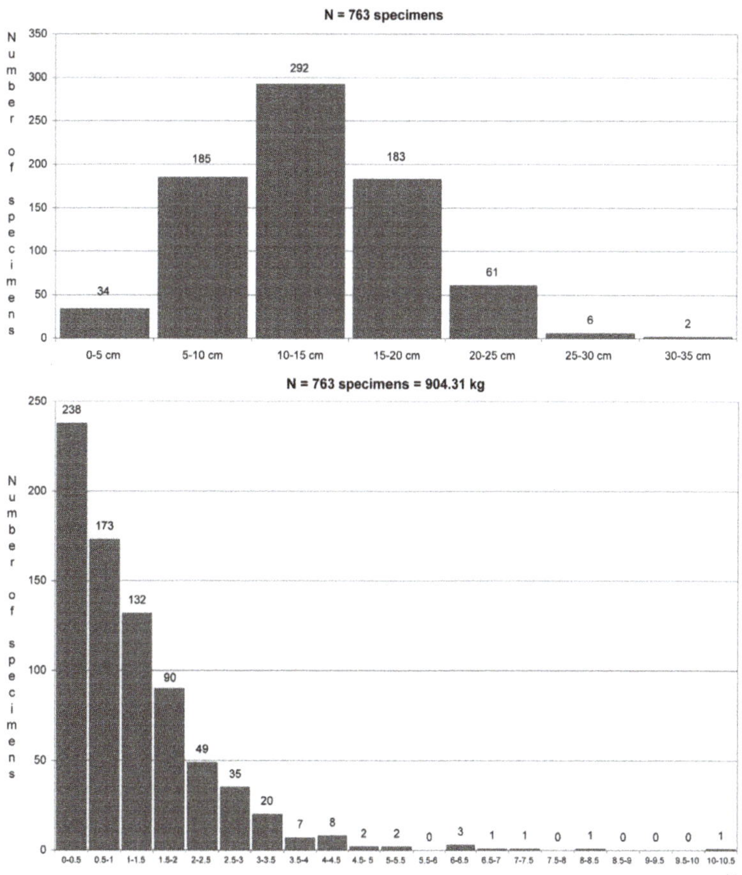

Fig. 30: Spiennes, Hainaut province. *Petit-Spiennes*. Shaft 20. Size and weight of flint nodules abandoned by miners in the exploitation level.

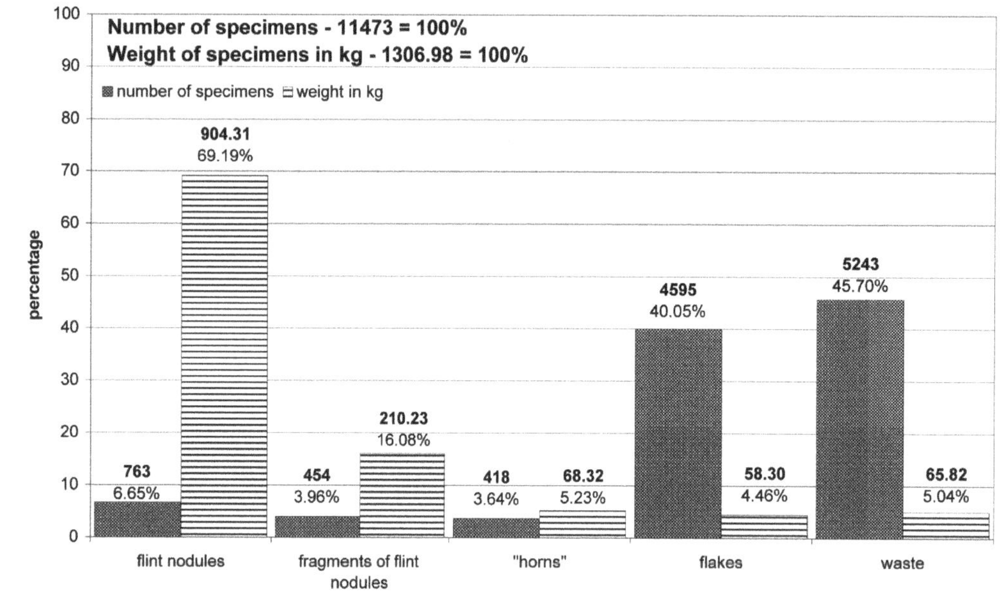

Fig. 31: Spiennes, Hainaut province. *Petit-Spiennes*. Shaft 20. Differentiation of basic categories of flint waste abandoned in the exploitation level, according to frequency and weight of specimens.

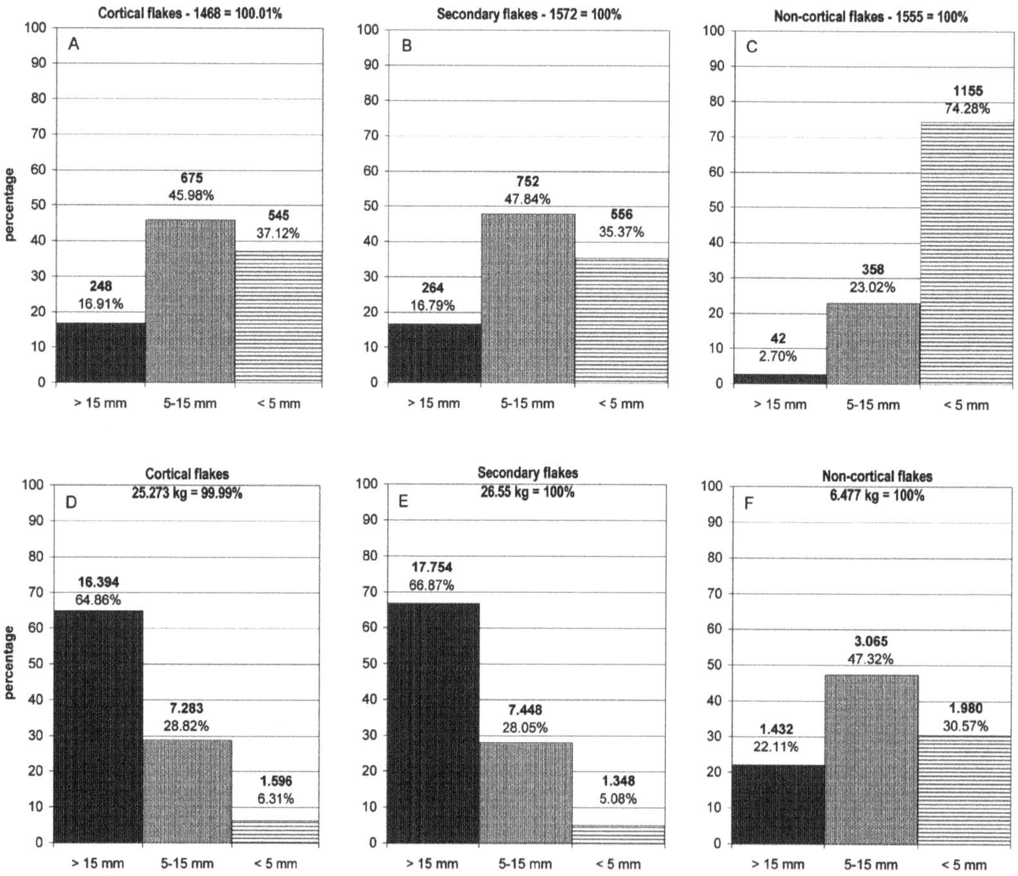

Fig. 32: Spiennes, Hainaut province. *Petit-Spiennes*. Shaft 20. Differentiation of basic categories of flakes abandoned in the exploitation level, according to thickness and upper surface of specimens.

Petit-Spiennes, constituting about 25 per cent of the total mass of dumps. The sample seems accurately to reflect the structure of the whole dump (Fig. 30-31). However, the analysis disregards microflakes and chips < 15 mm, together weighing 18.73 kg, and numerous flint picks which will be analysed in the future.
Interestingly, the sample did not contain flint hammers and only two sandstone hammers were found in the whole area explored in the shaft. Most of the hammers were probably removed from the shaft by the miners.

It is worth noting that many of the flint nodules extracted did not meet the miners' requirements, because of their size or shape, and were discarded straight away and left underground. A large number of such nodules were recovered – 763 specimens, constituting in weight over two thirds of all the flint left at the bottom (Fig. 30). The nodules were widespread throughout the dump, indicating that the nodule was assessed for suitability and, if needed, was disqualified immediately after extraction from the chalk. Over half of these discarded nodules weighed less than 1 kg. 38% were 10-15 cm in size and it is obvious that small specimens did not interest the miners; the same is probably true for the slightly bigger ones, of 15-20 cm (24%). We can assume that most of the nodules which were taken above ground were larger than 20 cm and weighed more than 2.5 kg (Fig. 30). The relatively high percentage of defective nodules means that the quality checks and the selection carried out at the bottom of the shaft were accurate. The relatively large percentage of nodule fragments - beside natural ones - indicates that damage also occurred during the extraction and the checking process.

At the same time as they were checked for quality, any kind of irregular protuberance was removed from the nodules. These 'horns' (418 pieces – 3.64%) were probably also perceived by the miners as diminishing the piece's worth (Fig. 30 and 32). Both procedures were complementary. Quality was also checked by striking off simple flakes, mainly cortical and secondary, and these two categories together predominate in number (66% of all flakes) and in weight (88.9%) (Fig. 33). Among the flint workshops from various European mines which have been studied so far, similar proportions have been determined at Sąspów, in the mine workshops located on the surface (Dzieduszycka-Machnikowa, Lech 1976: 92-100; Lech 1983: 52-64). The occurrence of non-cortical flakes resulting from the processing of the extracted nodules is in fact much lower than suggested by the diagram (Fig. 33: C and F), as part of the flakes in this category are the result of damage to and repair of picks. In the dump at the bottom of shaft 20, there is a large number of waste pieces.
To describe the flakes from the waste dump at the bottom of shaft 20, H. Collet used the method proposed by A. Augereau (1995) but supplemented it by counting the classified flakes (Fig. 33: A-C), though not the specimens in the chip category (<15 mm), which were only weighed. This allows the interesting possibility of comparing the number of pieces in the different categories of flakes with their weight and of evaluating the significance of the criterion of thickness, its correlation with the features of the dorsal face of the specimens and their weight (Fig. 33).

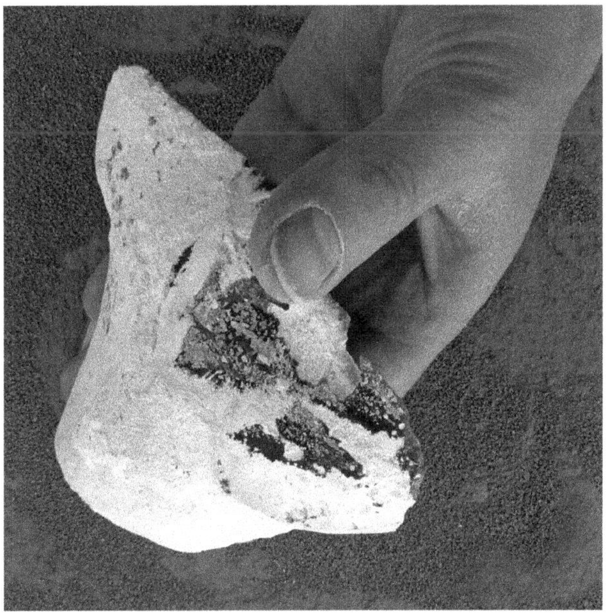

Fig. 33: Spiennes, Hainaut province. *Petit-Spiennes*. Shaft 20. A specimen from the 'horn' category abandoned in the exploitation level. Photograph by M. Woodbury.

The numbers of totally cortical, partially-cortical and non-cortical flakes in the dump at the bottom of shaft 20 were similar – 31.95%, 34.21% and 33.84% respectively, but in terms of weight, they were respectively: 43.35%, 45.54%, and 11.11% - so we see a clear difference in the mass of non-cortical flakes when compared with the other two groups. The differences in the thickness of flakes are even more interesting. Here the relation between the discrete categories of thickness shows clear differences and, moreover, the ratio of number to mass of flakes is inversely proportional. In number, flakes > 15 mm thick made up 12.06% of those found at the bottom of the shaft, flakes between 5 and 15 mm made up 38.84%, and flakes < 5 mm, 49.1%. In terms of mass the same categories constituted: 61.03%, 30.52% and 8.45%, so the infrequent very thick flakes were heaviest, while the most numerous category of thin flakes was the lightest. Thus we can regard the thickness of flakes as an important criterion for classification, in accordance with the postulates of A. Augereau (1995: 146) and J. Pelegrin (1995: 162), and in accordance with their place in the *chaîne opératoire*.

If we consider the differences in the flakes from the dump at the bottom of shaft 20 according to their connection with successive elements of the *chaîne opératoire*, 25.83% of the specimens, in regard to numbers, and 71.06%, in regard to weight, should be attributed to the stage of preparation of the nodule when very thick, and thick cortical flakes and very thick partially-cortical flakes, were detached by means of a hard hammer. This is an acceptable conclusion.

According to A. Augereau, flakes of medium thickness (5 to 15 mm), partially-cortical and non-cortical should be attributed to the stage when the nodule was being shaped using a hard or soft hammer. This would apply to 24.16% of the specimens with regard to number and 18.3% with regard to weight. The problem is that there are no indications that axes were prepared at the bottom of the shaft. It seems more likely that some of the flakes in this group are the result of incidental damage to the nodules when they were being extracted from the deposit with the use of picks. They could also have been the result of removal of flint protuberances from the nodule, the testing of the quality of the raw material and the forming and repairing of picks, or of damage to the picks sustained during work.

In contrast, most of the 25.14% of thin non-cortical flakes of ≤ 5 mm (probably small, considering their weight which was only 3.40% of all the flakes) and part of the uncounted small fraction of microflakes and chips of ≤ 15 mm and mass of 18.730 kg, were most probably the result of the final stage of preparing picks and other mining tools, of repairing them, and most certainly also of damage done them during work.

All in all, the criteria of thickness and weight proposed by A. Augereau seem pertinent, but there is no support for the assumption that in the case of flakes these criteria function more effectively than the criterion of size.

Concerning flint mining from the deepest shafts, such as shafts 1 and 2 at *Camp-à-Cayaux*, the organisation of work was completely different. The largest nodules were extracted there, from 1 to 2 m long, about 1 m wide and 15 to 30 cm thick. The nodules were split up underground and the raw material was brought up in the form of smaller blocks which were small and light enough to be transported from a considerable depth to the surface through narrow shafts. Probably part of the waste material was removed – natural pieces of the flint nodule, flint breccia, sparse detached flakes, etc.

Unfortunately, there are no observations which would help describe the behaviour of miners in the direct vicinity of the shafts. This is due to the fact that, at Spiennes, at no time were the dumps next to the mine shafts, or rather their modest remains preserved in some cases by the shafts, studied. Nor was there any research directed towards excavating the surface area surrounding the shafts which, in some cases, may have survived.

The raw materials were transported from the shafts to workshops located within the mining field. In workshops of the Michelsberg culture, blade cores were prepared and exploited, while part of the raw materials were directly transformed into flint axe blades; in some cases, (if suitable), used blade cores were turned into axes. At times, the numerous Spiennes picks were produced at the same time and in the same workshops as manufactured blade blanks and axe blades. This is true in the case of workshops II, III and IV, excavated by F. Hubert in 1965. Picks were prepared there from used blade cores (Hubert 1969: 27, 28 and 31). We know that in the Spiennes area there were also workshops where only axes were made, while no blades were produced.

Workshop productivity is difficult to determine. Only on the basis of F. Hubert's excavation from 1965 can we venture a guess that in the case of workshops I – IV the scale of production was not large. It probably comprised blades procured from about 100 cores, a similar number of axes and a substantially smaller number of picks. There is a large margin of error here. We cannot exclude the possibility that the same workshops also produced advanced pre-cores and prepared cores which, together with selected blades and flakes, roughouts of axe blades and initially prepared flint blocks, as raw material, were transported to the miners' own settlements or became goods for exchange with other communities. This possibility is indicated by the results of excavations at the settlement of a Michelsberg culture community in Thieusies (Vermeersch *et al.* 1990) at less than 15 km from the flint mines. Four flint picks found at the settlement suggest that the Thieusies community was directly engaged in exploiting the Spiennes flint. The large number of cores in Spiennes flint, as well as the presence of axe roughouts, also supports this hypothesis. Unfortunately, our great ignorance regarding settlements of the Michelsberg culture communities in Hainaut province, in all of Belgium and the neighbouring regions of France, makes further surmises as to the connections between the Spiennes mine complex and its settlement quite pointless.

9. THE *PETIT-SPIENNES* ENCLOSURE AND REMAINS OF DAILY LIFE

The Spiennes mining complex also included a settlement surrounded by a system of ditches and banks, considered by some authors as to have been a place serving some ceremonial function.

Discovered in the fifties, the enclosure is located on the *Petit-Spiennes* plateau south of the railway trench, on a site where some mining features were discovered, and in the immediate vicinity of the mining area of *Petit-Spiennes*. The enclosed area, labelled *Camp Michelsberg*, is situated near the top of the plateau and looks down upon the surrounding countryside. Archaeological excavations determined the location of ditches and only one related pit was excavated inside. As a result, a lot of flint flakes and some burnt daub, sandstone fragments, potsherds, burnt bones as well as charcoal were found.

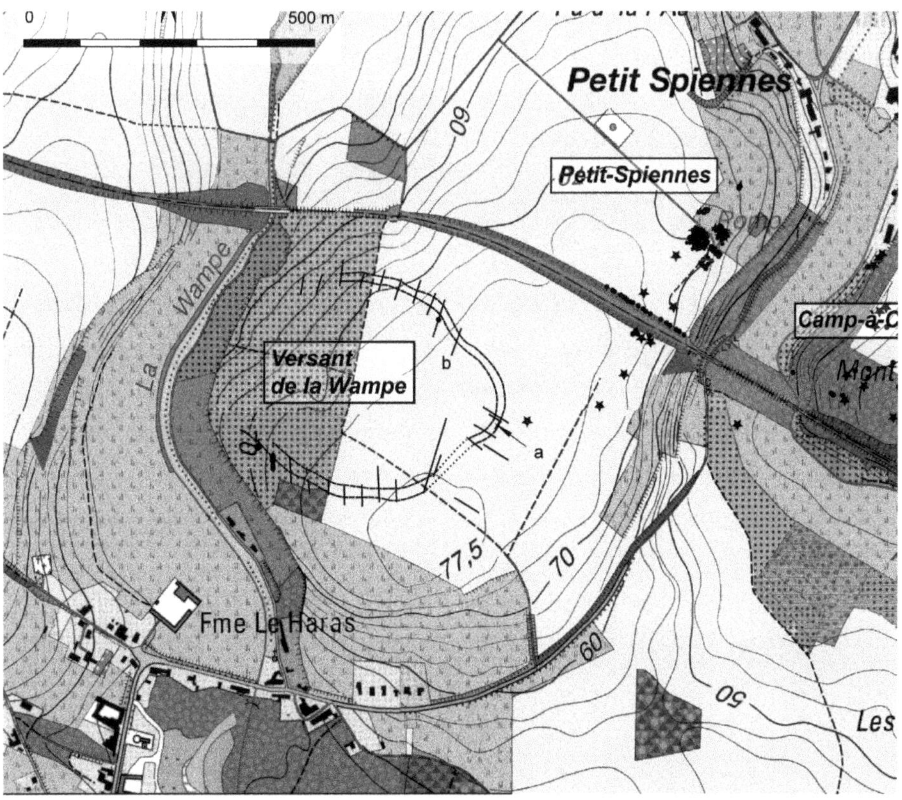

Fig. 34: Spiennes, Hainaut province. *Petit-Spiennes*. Michelsberg enclosure on the *Petit-Spiennes* plateau: a - interruption; b - excavation trenches. Drawing by M. Woodbury after F. Hubert (1971 and 1976a).

The layout of the ditches and the approximate size of the enclosure were determined only after some thirty trenches had been dug, though this was not enough to confirm whether the ditches were continuous or not. Only one attested interruption was discovered on the eastern part of the enclosure. For this reason, the drawing set out here (Fig. 34) gives only a general outline, based on an aerial view and on the excavation data published up to 1976 (Hubert 1971; 1976a).

The feature is composed of two parallel ditches 2.55 to 8 m wide and 1.3 to 2.45 m deep beneath the present day surface and enclosing an area of about 14 ha. The distance between the two ditches varies from 5 to 10 m. The presence, originally, of two banks on the inner side of each ditch was suggested by F. Hubert (1971), based on the distance between the two ditches and an examination of their fills.

Among the flint material from small excavation trenches, F. Hubert (1971: 42) distinguished flake and blade cores – the former in greater numbers. The most frequent tools were flake end-scrapers and axe roughouts; the latter were three times more numerous than polished axes. There were also chisels, side-scrapers, flint arrowheads, and a single burin and borer. The connection of the settlement with the mining and knapping of flint is supported by the presence of picks, hammers, polishing stones (*polissoirs*).

The pollen spectrum obtained from three samples taken in one profile, located in the deepest level of the inner ditch, shows that the enclosure was situated in an area which was probably largely cleared of forest. The arboreal pollen represents only 19 to 25 % of the total number of the pollen spectrum. It contrasts significantly with the results obtained for the mining field itself, which shows a greater quantity of trees and mainly bushes - from 42 to 95% (Defgnée, Collet 2003; Heim 2003). The human activity is also attested by the presence of cereals and ruderal plants which clearly indicate the presence of fields and grazing in the immediate vicinity (Heim 1971).

Unfortunately, no radiocarbon dates are, as yet, available. The archaeological material suggesting a date for the enclosure comes mainly from the ditches, especially from the entrance and the inner ditch. The pottery from the Middle Neolithic layers (it should be pointed out that material of later periods was found in the upper part of the ditches)

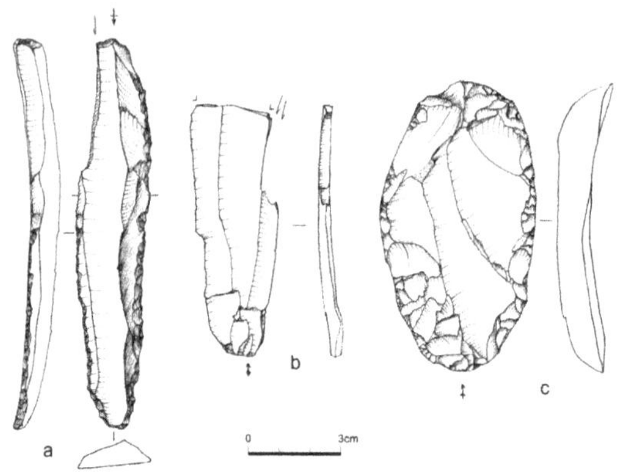

Fig. 35: Spiennes, Hainaut province. *Camp-à-Cayaux*. Flint implements. RBINS: Stevens collection. Fig. 35 a: drawing by A.-M. Wittek; Fig. 35 b-c: drawings by A. Hauzeur.

was first studied by F. Hubert (1971), and recently by B. Vanmontfort (2004). The archaeological material can be attributed to the Michelsberg culture (Hubert 1971) and, specifically, to the 'Central Scheldt basin group' (Vanmontfort 2004: 342). This group is connected with the Michelsberg culture groups from the Paris basin, the Middle Meuse and the Rhine basin. This material shows no real difference when compared with the pottery found in the mining areas: the same temper, same treatment, same shapes (Vanmontfort 2004: 214)... For this reason the settlement looks contemporaneous with at least a part of the mining activity at Spiennes.

Elsewhere, the Spiennes mining fields were dotted with structures connected with the exploitation and working of the flint. Excavations revealed Michelsberg potsherds, bones of consumed animals and 'domestic' flint tools, generally rare at Spiennes. End-scrapers, burins, blades and polished axes were found (Fig. 35). Unfortunately, again no clear interpretation can be given of these discoveries, which are generally old and insufficiently recorded.

10. HUMAN BONES REMAINS

It seems that most of the human bones discovered at Spiennes and in its vicinity give evidence of funerary practices from the Middle Neolithic period, other than those found during the same period in cave or rock-shelters in the Meuse basin (Cauwe *et al.* 2001).

Human remains are known from Obourg, Strépy and Spiennes. Some of these have been attributed to miners accidentally killed during their work, like the famous 'Obourg' miner. Some 15 years ago, most of these human bones were studied in detail, also in the historical context of their discovery, and partly dated (Heinzelin *et al.* 1993). In fact, some of these human remains turned out to be forgeries, dating from the turn of the 20th century. The forgers used human remains from the Merovingian period as well as protohistoric bones, as in the case of the 'miners' of Obourg and Strépy.

Nevertheless, the results of different analyses provided some information about the presence of human remains in a mining context. ^{14}C dates indicate that a skeleton from Spiennes has been correctly dated to the Middle Neolithic (Spiennes C; OxA-3196; Fig. 16). The most complete and certain find is a female adult with a perinatal child placed in the middle part of the filling sequence of shaft 11 at *Petit-Spiennes* (Toussaint *et al.* 1997; Collet, Toussaint 1998; Collet, Van Neer 2002). A ^{14}C date gives 4500 ± 50 BP (Beta-110683), which is within the range of dates for the flint mining at Spiennes (Fig. 16).

FINAL REMARKS

On the basis of the excavations carried out so far, it is difficult to determine how much mining and flint working was done overall in the Spiennes mining fields (Collet 2004: 132). We do not know the exact borders of the mining fields and therefore cannot calculate the number of mining features. There is also no precise dating for the various types of mining activities. Moreover, it is not known whether the flint mining at Spiennes was continuous, or whether the different phases of exploitation of the flint deposits were divided by time intervals when mining activity ceased.

The size of the Spiennes site complex and the advanced technology of the extraction process led some authors to infer that the mining and working of the flint there was highly specialised (Verheyleweghen 1966). Such conclusions need to be verified. For reasons mentioned earlier, we know little about the scale and extent of distribution of products from the workshops located next to the mines. According to available data and literature, there is no doubt that the blades and axe blades made in the mine workshops were distributed within a radius of 70 km from the mine, but this seems an exceedingly modest estimate.

As to the intriguing problem of labour specialisation, it seems much more likely that flint mining was just one of the spheres of activity of farming communities who exploited the mines and at the same time worked the land and bred animals (Gautier, Biondi 1993; Collet 2004: 132). These modest conclusions show the crucial problems which archaeology encounters to interpret such sites as Spiennes and other nearby mines. The key to solving them seems to lie in comparative studies of mining fields, flint workshops and flint material from settlements (Dzieduszycka-Machnikowa, Lech 1976; Saville 1981; Lech 1983; 1997: 624-628; 2004: 70-76; Augereau 1995; Cupillard *et al.* 1995; Jeudy *et al.* 1995; Pelegrin 1995; Lech, Longworth 2006).

Brussels – Warsaw – Mons – Paris, June 2005 – September 2007

Hélène Collet
Association sans but lucratif 'Société de Recherche préhistorique en Hainaut'
Maison Losseau
37, rue de Nimy
B-7000 Mons (Belgium)
e-mail: hcollet@tvcablenet.be

Dr Anne Hauzeur
INRAP Centre - Île-de-France, scientific collaborator at RBINS
148, Avenue André Maginot
F-37000 Tours (France)
e-mail: ahauzeur@yahoo.fr

Prof. Dr Jacek Lech
Polish Academy of Sciences
Institute of Archaeology and Ethnology
Aleja. Solidarności 105
PL 00-140 Warszawa (Poland)
e-mail: lech@iaepan.edu.pl

REFERENCES

Adam, A. 1959. Sur quelques outils à faciès d'habitat trouvés à Spiennes (Pa d'la l'iau). Annales du Cercle archéologique de Mons 63, 63-66, 4 pl.

Armstrong, A.L. 1923. Discovery of a new phase of early mining at Grime's Graves, Norfolk, Preliminary Report. Proceedings of the Prehistoric Society of East Anglia 4 (1), 113-125.

Armstrong, A.L.1927. The Grime's Graves problem in the light of recent researches. Proceedings of the Prehistoric Society of East Anglia 5(2), 91-136.

Armstrong, A.L.1934. The Grime's Graves, Norfolk. Report on the Excavation of Pit 12. Proceedings of the Prehistoric Society of East Anglia 7(3), 382-394.

Augereau, A.1995. Les ateliers de fabrication de haches de la minière du 'Grand Bois Marot ' à Villemaur-Sur-Vanne (Aube). In J. Pélegrin and A. Richard (eds.), Les mines de silex au Néolithique en Europe : avancées récentes, Actes de la table-ronde internationale de Vesoul, 18-19 octobre 1991, 145-158.

Bonenfant, P.-P. 1993. Spiennes (Mons, Hainaut): mines néolithiques de silex. Annales d'Histoire de l'Art et d'Archéologie de l'Université Libre de Bruxelles 15, 123-125.

Bonenfant, P.-P. and Auzou, G. 1993. Mons/Spiennes '*Camp-à-Cayaux*', fouilles de minières néolithiques. Campagne de 1992. Chronique de l'Archéologie en Wallonie 1, 22-23.

Borkowski, W., Migal, W., Sałaciński, S. and Zalewski, M. 1991. Possibilities of investigating Neolithic flint economies, as exemplified by the banded flint economy. Antiquity 65 (248), 607-627.

Bostyn, F. and Lanchon, Y. (eds.) 1992. Jablines: Le Haut Château (Seine-et-Marne). Une minière de silex au Néolithique. Paris: Editions de la Maison des Sciences de l'Homme, DAF 35, 246 pages.

Breuer, J. 1930. Recherches à Spiennes 3-20 mars 1930. Bulletin des Musées royaux d'Art et d'Histoire 3, 87-88.

Brézillon, M. 1971. La dénomination des objets de pierre taillée. Matériaux pour un vocabulaire des préhistoriens de langue française. IVe suppl. to Gallia Préhistoire. 2nd ed. Paris: Editions du CNRS.

Briart, A., Cornet, F.-L., Houzeau de Lehaie, A. 1868. Rapport sur les découvertes géologiques et archéologiques faites à Spiennes en 1867. Mémoires et publications de la Société des Sciences, des Arts et des Lettres du Hainaut, Troisième série, tome 2, année 1866-1867, 355-398.

Cartailhac, E. 1896. La France préhistorique d'après les sépultures et les monuments. 2nd ed,vol. I. Paris: Librairie Félix Alcan.

Cauwe, N., Vander Linden, M. and Vanmontfort, B. 2001. The Middle and Late Neolithic. In N. Cauwe, A. Hauzeur and P.-L.van Berg (eds.), Prehistory in Belgium. Special issue on the occasion of the XIVth Congress of the International Union for Prehistoric and Protohistoric Sciences. Anthropologica et Praehistorica 112, 77-89, 156-163.

Cels, A. and De Pauw, L. 1885-1886. Considérations sur la taille du silex telle qu'elle était pratiquée à Spiennes à l'âge de la pierre polie. Bulletin de la Société d'Anthropologie de Bruxelles 4, 245-246.

Centre archéologique de Recherche minière (ed.) 1990. Etude préalable à la sauvegarde et à la mise en valeur du site archéologique de Spiennes - Mons, Spiennes.

Childe, V.G. 1925. The Dawn of European Civilization. 1st ed. London: Routledge, and Kegan Paul Ltd.

Childe, V.G. 1957. The Dawn of European Civilization. 6th ed. London: Routledge, and Kegan Paul Ltd.

Clark, J.G.D. 1952. Prehistoric Europe. The Economic Basis. London: Methuen, and Co. Ltd.

Clark, G. 1977. World prehistory in new perspective. Cambridge: Cambridge University Press.

Clark, G. and Piggott, S. 1933. The Age of the British Flint Mines. Antiquity 7, 166-183.

Clarke, W.G. (ed.) 1915. Report on the Excavations at Grime's Graves, Weeting, Norfolk, March-May, 1914. London: Prehistoric Society of East Anglia.

Collet, H. 2000. Fouille de nouveaux puits d'extraction à *Petit-Spiennes* et découverte de fragments de polissoir. Notae Praehistoricae 20, 163-170.

Collet, H. 2003a. Mons/Spiennes: indices d'excavations souterraines dans les parcelles 406c et 407a de '*Petit-Spiennes*'. Chronique de l'Archéologie wallonne 11, 40-42.

Collet, H. 2003b. Mons/Spiennes: fouille dans la parcelle 393c de '*Petit-Spiennes*'. Chronique de l'Archéologie wallonne 11, 39-40.

Collet, H. 2004. Les mines néolithiques de Spiennes: état des connaissances et perspectives de recherche. In Section 9. The Neolithic in Near East and Europe. Acts of the XIVth UISPP Congress, University of Liège, Belgium, 2-8 September 2001, BAR International Series 1303, Oxford: BAR, 129-133.

Collet, H., Deramaix, I., Sartieaux, P. and Vander Linden, M. 1997. Fouille préventive de puits d'extraction de silex à *Petit-Spiennes* (Hainaut). Notae Praehistoricae 17, 203-212.

Collet, H. and Toussaint, M. 1998. Découverte d'un squelette humain néolithique sur le site minier de *Petit-Spiennes* (Hainaut, Belgique): étude préliminaire. Internéo 2. Journée d'information du 14 novembre 1998, Paris, 113-124.

Collet, H. and Vander Linden, M. 1998. *Petit-Spiennes*: compte-rendu de la campagne 1998. Notae Praehistoricae 18, 167-173.

Collet, H. and Vankerkhoven, A. 2004. Mons/Spiennes: poursuite des fouilles dans la parcelle 393c de '*Petit-Spiennes*'. Chronique de l'Archéologie wallonne 12, 39-40.

Collet, H. and Van Neer, W. 2002. Stratigraphie et faune d'un puits d'extraction néolithique à *Petit-Spiennes*. Anthropologica et Prehistorica 113, 73-104.

Collet, H. and Woodbury, M. 2000. Mons/Spiennes: fouille de puits d'extraction de silex à *Petit-Spiennes*, Chronique de l'Archéologie wallonne 8, 30-31.

Collet, H. and Woodbury, M. 2001. Mons/Spiennes: poursuite de la fouille de puits d'extraction de silex au lieu-dit '*Petit-Spiennes*'. Chronique de l'Archéologie wallonne 9, 39-40.

Collet, H. and Woodbury, M. 2002. Mons/Spiennes: campagne de fouille de puits d'extraction de silex au lieu-dit '*Petit-Spiennes*'. Chronique de l'Archéologie wallonne 10, 37-38.

Collet, H. and Woodbury, M. 2006. Mons/Spiennes: poursuite des fouilles dans la parcelle 393c de '*Petit-Spiennes*'. Chronique de l'Archéologie wallonne 13, 47-48.

Collet, H. and Woodbury, M. in press a. Mons/Spiennes: découverte d'ateliers et de puits d'extraction du silex rue d'Harmignies. Chronique de l'Archéologie wallonne 14.

Collet, H. and Woodbury, M. in press b. Mons/Spiennes: indices miniers néolithiques dans la parcelle 169a au '*Camp-à-Cayaux*'. Chronique de l'Archéologie wallonne 14.

Collet, H. and Woodbury, M. in press c. Mons/Spiennes: découverte de vestiges d'époque indéterminée rue du Moulin. Chronique de l'Archéologie wallonne 14.

Colman, P. 1957. Le Néolithique et ses prolongements à Spiennes. Bulletin de la Société royale belge d'Etudes géologiques et archéologiques 'Les Chercheurs de la Wallonie' 16, 226-290.

Cupillard, Ch., Affolter, J. with Campy M., Contini, D. and Richard, H. 1995. La minière de silex néolithique de Blanc-Saule à Etrelles-et-la-Montbleuse (70) et l'exploitation du silex lacustre oligocène inférieur de Haute-Saône durant le Néolithique, in: J. Pelegrin and A. Richard (eds.), Les mines de silex au Néolithique en Europe : avancées récentes. Actes de la table-ronde internationale de Vesoul, 18-19 octobre 1991, 179-240.

Déchelette, J. 1908. Manuel d'Archéologie préhistorique, celtique et gallo-romaine I, Paris.

Defgnée, A. and Collet, H. 2003. Mons/Spiennes: premiers résultats palynologiques dans la zone minière de '*Petit-Spiennes*'. Chronique de l'Archéologie wallonne 11, 43-44.

De Koninck, M. 1860. Sur quelques ossements humains fossiles et sur quelques silex taillés. Notice de M. C. Malaise. Rapport de M. De Koninck. Bulletin de l'Académie royale des Sciences, des Lettres et des Beaux-arts de Belgique 10 (29ème année, 2ème série), 511-514.

De Laet, S.J. 1982. La Belgique d'avant les Romains. Wetteren: Universa.

Delvaux, E. 1885-1886. Excursion de la société à Mesvin, à Spiennes et à Harmignies le 5 septembre 1885. Bulletin de la Société d'Anthropologie de Bruxelles 4, 176-208.

De Pauw, L. and van Overloop, E. 1889-1890. Les ateliers préhistoriques de Spiennes. Bulletin de la Société d'Anthropologie de Bruxelles 8, 28-39.

Desterbecq, D. 2004. Mons/Spiennes: fouilles de minières néolithiques dans la parcelle 406b de '*Petit-Spiennes*'. Chronique de l'Archéologie wallonne 12, 38-39.

Desterbecq, D. and Joris, J.-P. 2006. Mons/Spiennes: fouille de minières néolithiques à '*Petit-Spiennes*', Chronique de l'Archéologie wallonne, 13, 44-45.

Di Lernia, S., Fiorentino, G., Galiberti, A. and Basili, R. 1995. The Early Neolithic mine of Defensola 'A' (I 18): flint exploitation in the Gargano area. Archaeologia Polona 33, 119-132.

Dzieduszycka-Machnikowa, A. and Lech, J. 1976. Neolityczne zespoły pracowniane z kopalni krzemienia w Sąspowie. (Polskie badania archeologiczne 19). Wrocław-Warszawa-Kraków-Gdańsk: Zakład Narodowy imienia Ossolińskich.

Evans, J. 1872. The Ancient Stone Implements, Weapons and Ornaments, of Great Britain. London: Longmans, Green, Reader, and Dyer.

Fechner, K., Gillet, E., Henton, A., Laurent, C., Mathieu, S. and Van Assche, M. 1993. Étude archéologique du tracé du gazoduc Flobecq-Quévy, Ministère de la Région wallonne, Convention n° 9311610, unpublished report.

Felder, P.J., Rademakers, P.C.M. and De Grooth, M.E.Th. (eds.) 1998. Excavations of Prehistoric Flint Mines at Rijckholt-St. Geertruid (Limburg, The Netherlands) by the 'Prehistoric Flint Mines Working Group' of the Dutch Geological Society, Limburg Section, Bonn: Deutschen Gesellschaft für Ur- und Frühgeschichte e.V. in Kommission bei Dr. Rudolf Habelt GmbH, (Archäologische Berichte 12).

Gautier, A. and Biondi, S. 1993. Forgotten Faunal Collection from the Prehistoric Flint Mines at Spiennes (Province of Hainaut, Belgium). Helinium 33 (1), 80-98.

Gosselin, F. 1986. Un site d'exploitation du silex à Spiennes (Hainaut), au lieu-dit '*Petit-Spiennes*'. Vie archéologique 22, 33-160.

Grooth, M.E.Th. de 1991. Socio-economic aspects of Neolithic mining: a preliminary study. Helinium, 31(2), 153-189.

Grooth, M.E.Th. de 1997. The social context of Neolithic flint mining in Europe. In R. Schild and Z.Sulgostowska (eds.), Man and Flint, Warszawa: Institute of Archaeology and Ethnology, Polish Academy of Sciences, 71-75.

Hamal-Nandrin, J. and Servais, J. 1925. Compte-rendu de fouilles entreprises dans des emplacements d'habitations et d'ateliers néolithiques à Spiennes (près de Mons, province de Hainaut, Belgique). Bulletin de la Société préhistorique française, 22, 73-103.

Hawkes, J. 1965. History of Mankind: cultural and scientific development. Volume One, Part I: Prehistory. New York: The New American Library, Inc.

Heim, J. 1971. Datation par la palynologie d'un fond de fossé de la civilisation Michelsberg à *Petit-Spiennes*. In F. Hubert, Fossés néolithiques à Spiennes. Premier rapport. (Archaeologia Belgica 136), Bruxelles: Service national des Fouilles, 62-66.

Heim, J., Collet, H., Woodbury, M., Vankerkhoven, A. and Jadin, I. 2003. Etude palynologique de structures néolithiques fouillées à *Petit-Spiennes* en 1987, Notae Praehistoricae 23, 141-145.

Heinzelin, J. de, Orban, R., Roels, D. and Hurt, V. 1993. Ossements humains dits néolithiques de la région de Mons (Belgique), une évaluation. Bulletin de l'Institut royal des Sciences naturelles de Belgique, Sciences de la Terre 63, 311-336.

Hensel, W. and Tabaczyński, S. 1978. Rewolucja neolityczna i jej znaczenie dla rozwoju kultury europejskiej. Wrocław: Zakład Narodowy imienia Ossolińskich.

Hubert, F. 1969. Fouilles au site minier néolithique de Spiennes. Campagne de 1965. (Archaeologia Belgica 111). Bruxelles: Service national des Fouilles.

Hubert, F. 1971. Fossés néolithiques à Spiennes. Premier rapport. Annexe de J. Heim, (Archaeologia Belgica 136). Bruxelles: Service national des Fouilles.

Hubert, F. 1976a. Puits de mines à la tranchée du chemin de fer à Spiennes. Archaeologia Belgica 186, Conspectus MCMLXXV, 9-11.

Hubert, F. 1976b. Une nouvelle zone minière à Spiennes. Archaeologia Belgica 186, Conspectus MCMLXXV, 12-15.

Hubert, F. 1976c. Atelier de taille du silex de l'Âge du Bronze à Spiennes. Archaeologia Belgica 186, Conspectus MCMLXXV, 16-20.

Hubert, F. 1978. Une minière néolithique à silex au *Camp-à-Cayaux* de Spiennes. (Archaeologia Belgica 210), Bruxelles: Service national des Fouilles.

Hubert, F. 1980. Zum Silexbergbau von Spiennes (B 1). In G. Weisgerber, R. Slotta, and J. Weiner (eds.), 5000 Jahre Feuersteinbergbau. Die Suche nach dem Stahl der Steinzeit, Bochum: Deutschen Bergbau-Museum, 124-139.

Hubert, F. 1983. Fouille au puits n°2 du *Camp-à-Cayaux* de Spiennes (Hainaut), Archaeologia Belgica 253, 17-21.

Hubert, F. and Soumoy, M. 1993. Mons/Spiennes: fouille de prévention avant l'aménagement de la Station de Recherches, Chronique de l'Archéologie en Wallonie 1, 22.

Jażdżewski, K. 1984. Urgeschichte Mitteleuropas, Wrocław: Zakład Narodowy imienia Ossolińskich.

Jeudy, F., Jeunesse, Ch., Monnier, J.-L., Pelegrin, J., Petrequin, A.-M., Petrequin, P. and Praud, I. 1995. Les carrières néolithiques de Plancher-Les-Mines (Haute-Saône). Exemples d'une approche intégrée. In J. Pelegrin and A. Richard (eds.), Les mines de silex au Néolithique en Europe: avancées récentes. Actes de la table-ronde internationale de Vesoul, 18-19 octobre 1991, 241-280.

Krukowski, S. 1939. Krzemionki Opatowskie. Warszawa: Muzeum Techniki i Przemysłu.

Lech, J. 1981. Flint mining among the early farming communities of central Europe, Przegląd Archeologiczny, (Wrocław) 28, 5-55.

Lech, J. 1982-1983. Flint work of the early farmers. Production trends in central European chipping industries from 4500-1200 BC. An outline. Acta Archaeologica Carpathica (Kraków) 22, 5-63.

Lech, J. 1983. Flint mining among the early farming communities of central Europe. Part II - Basis of research into flint workshops. Przegląd Archeologiczny (Wrocław) 30, 47-80.

Lech, J. 1991. The Neolithic-Eneolithic transition in prehistoric mining and siliceous rock distribution. In J. Lichardus (ed.), Die Kupferzeit als historische Epoche, Part 2, Bonn: Dr. Rudolf Habelt GMbH, 557-574.

Lech, J. 1997. Remarks on Prehistoric Flint Mining and Flint Supply in European Archaeology. In A. Ramos-Millán and M. A. Bustillo (eds.), Siliceous Rocks and Culture, Granada: Editorial Universidad de Granada, 611-637.

Lech, J. 2004. O badaniach prehistorycznego górnictwa krzemienia i kopalni w Krzemionkach Opatowskich. Przegląd Archeologiczny (Wrocław) 52, 15-88.

Lech, J. and Leligdowicz, A.1998. Der Silexbergbau. In J. Preuß (ed.), Das Neolithikum in Mitteleuropa: Kulturen – Wirtschaft – Umwelt vom 6. bis 3 Jahrtausend v.u.Z., Weissbach, Beier and Beran (Archäologische Fachliteratur, vol. 1/1), 121-136.

Lech, J. and Longworth, I. 2000. Kopalnia krzemienia Grimes Graves w świetle nowych badań. Przegląd Archeologiczny (Wrocław) 48, 19-73.

Lech, J. and Longworth, I. 2006. The Grimes Graves flint mine site in the light of two Late Neolithic workshop assemblages: a second approach. In G. Körlin and G. Weisgerber (eds.), Stone Age - Mining Age, Der Anschnitt 19, 413-422.

Lefort, M. and Cody, G. 1953. Les cahiers de Spiennes, Octobre - Décembre 1953, 2nd ed. fasc. 1.

Lefort, M. 1954. Les cahiers de Spiennes. Spiennes. Février - mars 1954, fasc. 2.

Lichardus, J., Lichardus-Itten, M., Bailloud, G., Cauvin, J. *et al.* 1985. La Protohistoire de l'Europe. Le Néolithique et le Chalcolithique entre la Méditerranée et la mer Baltique. Paris: Presses Universitaires de France.

Loë, A. de 1913. Les fouilles de M. Louis Cavens à Spiennes en 1912. Bulletin des Musées royaux du Cinquantenaire 12, 36-37 and 44-46.

Loë, A. de 1914. Les fouilles de M. Louis Cavens à Spiennes en 1913. Bulletin des Musées royaux du Cinquantenaire 13, 35-37.

Loë, A. de 1925. Notice sur les fouilles exécutées à Spiennes en 1912, 1913 et 1914. Bulletin de la Société d'Anthropologie de Bruxelles 40, 151-171.

Loë, A. de and Munck, E. de 1891. Notice sur des fouilles pratiquées récemment sur l'emplacement du vaste atelier néolithique de Spiennes (Hainaut). Section: Ateliers et puits d'extraction de silex en Belgique, en France, en Portugal, en Amérique. In Congrès international d'Anthropologie et d'Archéologie préhistoriques. Compte-rendu de la dixième session (Paris: 1889), Paris, 569-602.

Loë, A. de and Rahir, E. 1929. Notice sur les fouilles exécutées à Spiennes en 1925 et en 1928. Bulletin de la Société d'Anthropologie de Bruxelles 44, 52-69.

Longworth, I., Herne, A. Varndell, G. and Needham, S. 1991. Excavations at Grimes Graves, Norfolk, 1972-1976. Fascicule 3. Shaft X: Bronze Age Flint, Chalk and Metal Working. London: British Museum Press.

Longworth, I. and Varndell, G. 1996. Excavations at Grimes Graves, Norfolk, 1972-1976. Fascicule 5. Mining in the Deeper Mines, London: British Museum Press.

Mariën, M.-E. 1961. La période de La Tène en Belgique. Le groupe de la Haine. (Monographies d'Archéologie Nationale 2), 92-132. Bruxelles: Musées royaux d'Art et d'Histoire.

Mercer, R.J. 1981. Grimes Graves, Norfolk. Excavations 1971-72: Volume I and II. London: Her Majesty's Stationery Office.

Milisauskas, S. and Kruk, J. 2002. Middle Neolithic: continuity, diversity, innovations, and greater complexity, 5500/5000-3500/3000 BC. In S. Milisauskas (ed.), European Prehistory. A Survey, 193-246. New York: Kluwer Academic/Plenum Publishers.

Mortillet, G. de 1885. Le préhistorique, antiquité de l'homme. 2nd ed. Paris: C. Reinwald.

Pelegrin, J. 1995. Réflexions méthodologiques sur l'étude de séries lithiques en contexte d'atelier ou de mine. In J. Pelegrin and A. Richard (eds.), Les mines de silex au Néolithique en Europe: avancées récentes. Actes de la table-ronde internationale de Vesoul, 18-19 octobre 1991, 159-165.

Piggott, S. 1954. The Neolithic Cultures of the British Isles, Cambridge: Cambridge University Press.

Robaszynski, F. and Christensen, W.K. 1989. The upper Campanian-Lower Maastrichtian chalks of the Mons basin, Belgium: a preliminary study of belemnites and foraminifera in the Harmignies and Ciply areas. Geologie en Mijnbouw 68, 391-408.

Robert, C. 2003. Mons/Spiennes : les minières néolithiques de '*Petit-Spiennes*'. Chronique de l'Archéologie wallonne 11, 42-43.

Saville, A. 1981. The Flint Assemblage. In R. J. Mercer, Grimes Graves, Norfolk. Excavations 1971-7, Volume II, 1-178. London: Her Majesty's Stationery Office.

Scollar, I. 1955. A Neolithic Enclosure at Spiennes. Antiquity 29 (115), 159-161.

Tabaczyński, S. 1970. Neolit środkowoeuropejski: podstawy gospodarcze. Wrocław: Zakład Narodowy imienia Ossolińskich.

Tabaczyński, S. 1972. Gesellschaftsordnung und Güteraustausch im Neolithikum Mitteleuropas. In B. Thaler (ed.), Neolithische Studien I, Berlin: Akademie-Verlag, (Wissenschaftliche Beiträge der Martin-Luther-Universität Halle – Wittenberg 1972/1, L 7), 31-96.

Toilliez, D. 1851. Troisième notice sur des antiquités découvertes dans le Hainaut. Bulletin de l'Académie royale de Belgique 18 (1ère partie), 659-670.

Toussaint, M., Collet, H. and Vander Linden, M. 1997. Découverte d'un squelette humain dans le puits de mine néolithique ST 11 de *Petit-Spiennes* (Hainaut). Notae Praehistoricae 17, 213-219.

Vandevelde, P. and Hubert, F. 1987. Deux vases Michelsberg de Spiennes et le matériel associé. Bulletin de la Société royale belge d'Anthropologie et de Préhistoire 98, 223-234.

Vanmontfort, B. 2004. Converging Worlds. The Neolithisation of the Scheldt basin during the late fifth and early fourth millennium calBC. Unpublished PhD. Leuven: Katolieke Universiteit Leuven, Department Archeologie, Kunstwetenschappen en Musicologie, Afdeling Archeologie.

Vanmontfort, B., Collet, H. and Crombé, Ph. (in press). Les Industries lithiques taillées des 4ème et 3ème millénaires dans les bassins de l'Escaut et de la Meuse (Belgique). In Actes du colloque 'Les industries lithiques taillées des 4ème et 3ème millénaires en Europe occidentale', Toulouse, 6-9 avril 2005.

Verheyleweghen, J. 1953. Découverte sur le territoire de Spiennes d'une phase d'occupation des hommes de Michelsberg, antérieure à celle du plateau du '*Camp-à-Cayaux*' (Compte-rendu des fouilles entreprises en 1953). Bulletin de la Société royale belge d'Anthropologie et de Préhistoire 64, 141-162.

Verheyleweghen, J. 1962. Un dépôt funéraire de crâne néolithique à Spiennes (Hainaut). Avec une note concernant l'importance de la population spiennienne et un essai sur la structure sociale de la tribu. Helinium 2 (3), 193-214.

Verheyleweghen, J. 1963. Evolution chronologique du Néolithique au '*Camp-à-Cayaux*' de Spiennes (Hainaut) avec une étude statistique dimensionnelle et caractérielle du matériel lithique. Helinium 3 (1), 3-38.

Verheyleweghen, J. 1966. Le Néolithique minier belge, son origine et ses relations culturelles. Second Colloque. Palaeohistoria 12, Acta et Communicationes Instituti Bio-Archaeologici universitatis Groningae. Proceedings of the Second Atlantic Colloquium (Groningen 6-11 April 1964), 529-557.

Vermeersch, P.M., Vynckier, G. and Walter, R. 1990. Thieusies Ferme de l'Hosté, site Michelsberg. II Le matériel lithique. (Studia Praehistorica Belgica 6). Leuven: KULeuven.

Weisgerber, G., Slotta, R. and Weiner, J. (eds.) 1980. 5000 Jahre Feuersteinbergbau. Die Suche nach dem Stahl der Steinzeit. Bochum: Deutschen Bergbau-Museum.

Whittle, A. 1996. Europe in the Neolithic. The creation of new worlds. Cambridge: Cambridge University Press.

Zawisza, J. 1873. Kongres Antropologii i Archeologii Przedhistorycznej, w Bruxelli 1872 r. Warszawa.

ACKNOWLEDGMENTS

The study of the material from the Stevens collection kept at the Royal Belgian Institute of Natural Sciences was granted by the Synthesys BE-TAF programme under the reference 1055.

A NEW FLINT MINE AT FLINS-SUR-SEINE/ AUBERGENVILLE (YVELINES, FRANCE)

Françoise Bostyn, François Giligny, Adrienne Lo Carmine

Abstract: Discovered in the 1920s, and located in the Seine valley downstream from Paris, the mine of Flins-sur-Seine/ Aubergenville was identified as a flint mine from recent aerial photographs. The site of Flins-sur-Seine exploited Bartonian Tertiary flint to manufacture axes. Surface collection and geophysical surveys, underway since 2002, show that the area with knapping waste covers about 15 hectares and that the extraction zone extends over at least 3 to 5 hectares. Evidence of first technological stages of axe manufacture, such as bifacial preparation, roughouts and knapped axes/axe-cores are present, but neither polished artefacts nor traces of settlement were found.

Keywords: Flins, flint mine, Bartonian flint, geophysical surveying, axe production.

1. LOCALISATION OF THE MINE AND RESEARCH HISTORY

The mine of Flins-sur-Seine / Aubergenville, to the West of Paris, dominates the Seine valley on the right bank of one of its tributaries, the Mauldre. The mine sits atop a plateau at an altitude of between 100 and 125 m, along a slight southeast / northwest slope (Fig. 1a). In this area, the Seine valley, as well as the secondary valleys of the Mauldre, and Vaucouleurs, cut deeply into the Tertiary geological substratum (Eocene, Oligocene) as far as the secondary formations (Cretaceous). White chalk with flint nodules lies at the base of the stratigraphic sequence, which is covered over by Ypresian and Lutetian layers. Limestone and marl of the Upper Bartonien (Ludien) or Lower (Saint-Ouen limestone) complete the stratigraphy. Siliceous material is abundant and varied, and the Saint-Ouen limestone is known for its brown to light beige flint found in slabs or nodules (Fig. 1b).

One of the earliest references to the mine of Flins-sur-Seine dates from 1926, with the work of S. de Sacy and M. Baudouin concerning the location, characterization and cultural attribution of items from the small hamlet of 'Le Clos' (De Sacy and Baudouin, 1926 and 1927). The variety of axes and the small number of polished axes prompted them to attribute this material to the end of the 'Campignien' (terminology of the time). These authors also noted the resemblance of the material at Flins with that of the Jablines, classifying Flins-sur-Seine in the 'upper Campignien', or 'Jablinien' (now obsolete terminology).

The site has been the object of several surface surveys, some of which have been published or officially reported (Daniel, 1965; Bourgeois, 1982). The mine has been regularly surveyed since the 1980's by many enthusiasts who have, over the years, made up impressive collections both in quantity and quality, but the site has never been excavated. As part of a collective research programme begun in 2000, we have undertaken a synthesis of this wealth of data by through two complementary approaches: mapping the site by using different surveying methods, and re-examining the artefacts and the historical data from the perspective of the technologies of production.

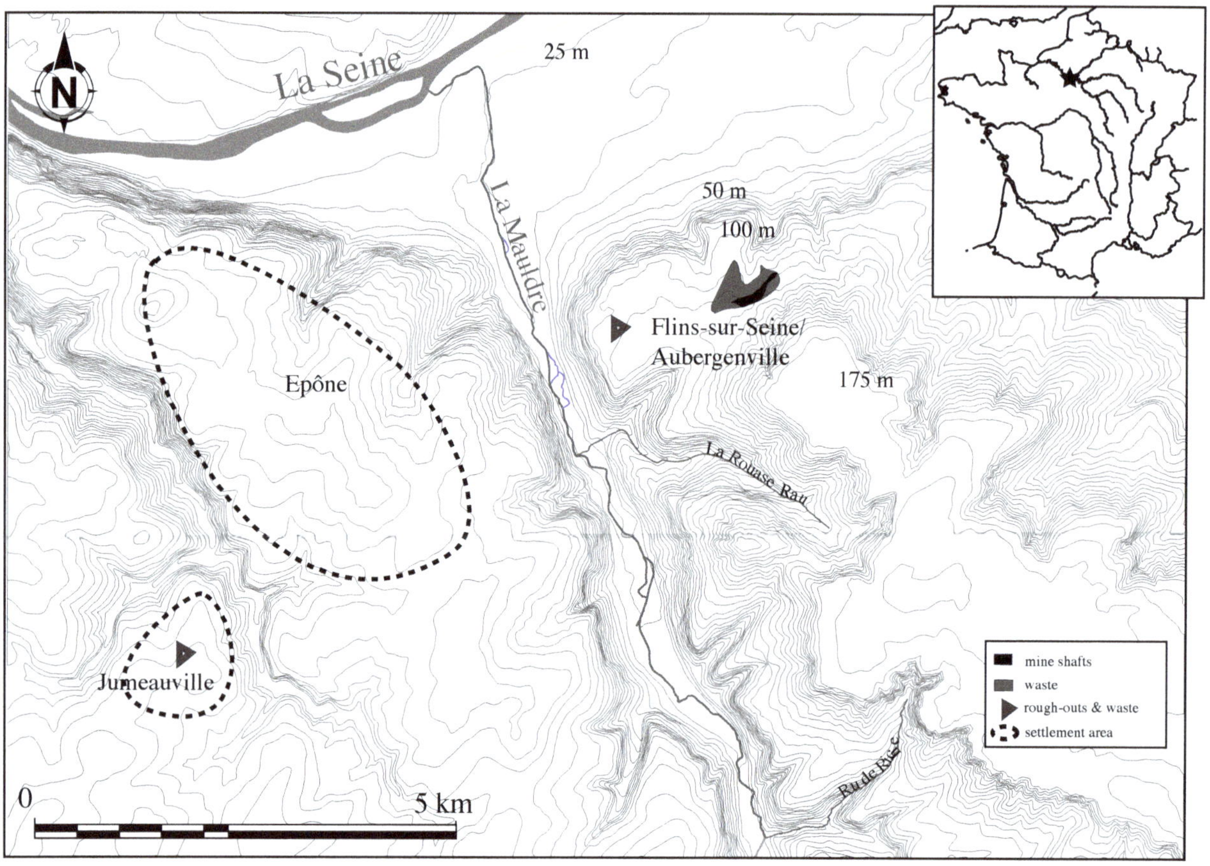

Fig. 1: A. Location of the mine Flins-sur-Seine/Aubergenville. K. Larsen del. 1999

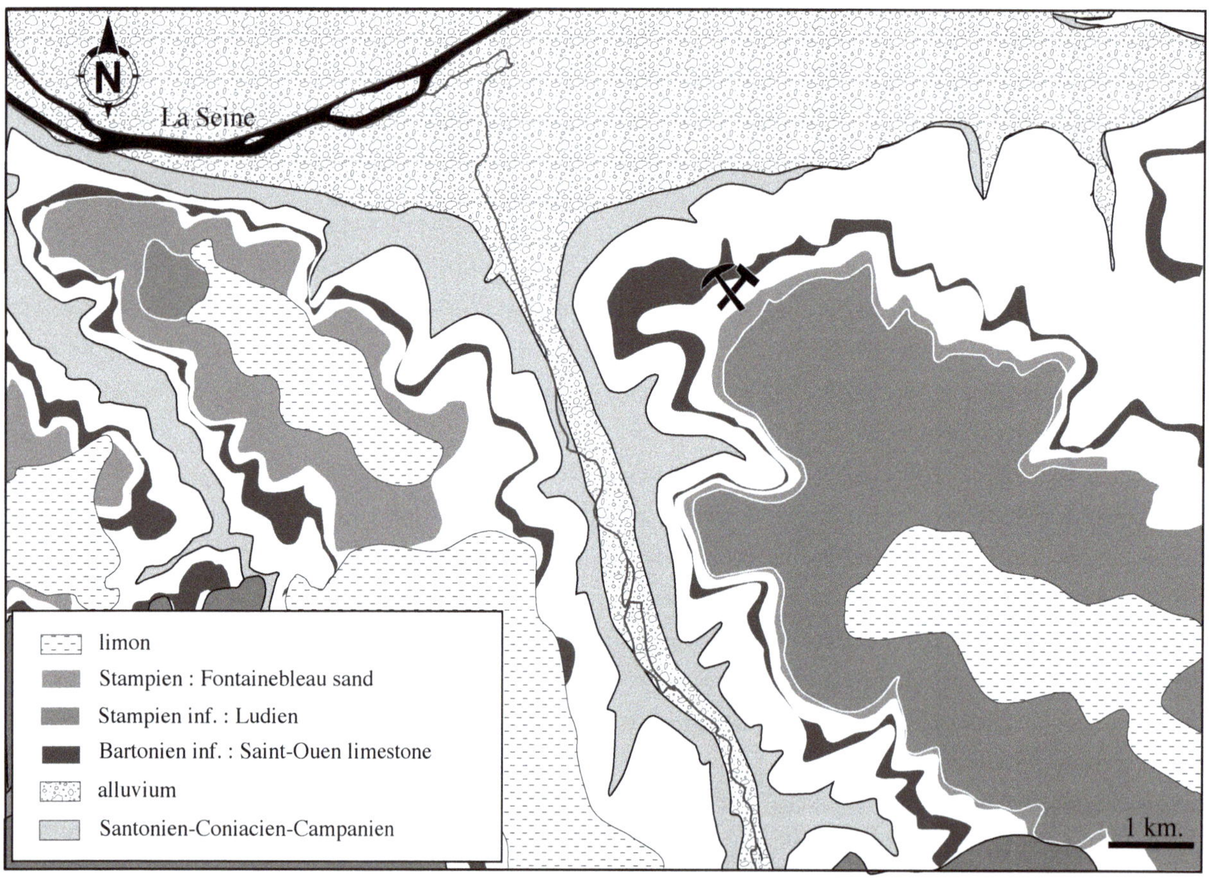

Fig. 1: B. Geological map (BRGM).

2. RECENT FIELD DATA (2000-2003)

Since 2000, different surveying methods have been gradually brought into play. Their continuous interaction has so far produced relevant results in different sectors of the site. The objectives of the overall survey are to establish the limits of the mining areas in relation to the size of the flint products.

2.1. Aerial surveying

In order to make an archaeological map, aerial photographs of the mine were taken by the Archaeological Service of the Department of Les Yvelines in 1999. Adjusted and placed on cadastral surveys of the area, the photographs allowed us to identify the presence of mine shafts in certain areas (work done by Terra NovA: 2001). Three concentrations appear from east to the west. Based on anomalies visible in aerial photography, interpreted as mining structures, an estimate of the number of such structures was made in the areas of interest. Two test areas were chosen for further investigation: one in the central part, the second in the eastern part of the site.

The central point of the anomalies was pinpointed and its coordinates recorded in a GIS in order to make an accurate count of the anomalies to calculate the average distance between the structures (Fig. 2). Within the two areas, the

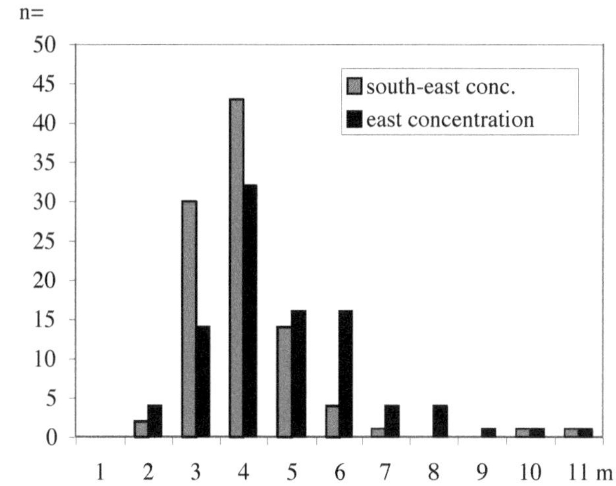

Fig. 3: Distribution of the distances between the anomalies identified on aerial photographs (pit) for two concentrations (metres).

first of 0.37 ha and the second of 1.5 ha, 96 anomalies were recorded. Distances between the structures are slight, in most cases less than 5 m. The average distance between the structures in the Eastern zone is 4.3 m, while this average is 3.5 m in the central area. Comparing distribution histograms of the distances for the two areas shows a greater coefficient of variation in the Eastern zone, structurally less dense than the other (Fig. 3). Almost one third of the distances in the Central zone are smaller than 3 m, compared to one fifth for those in the Eastern zone. However, the Eastern zone presents a fairly large number of distances greater than 5 m (29%), a value which is significantly smaller in the other sector (7%).

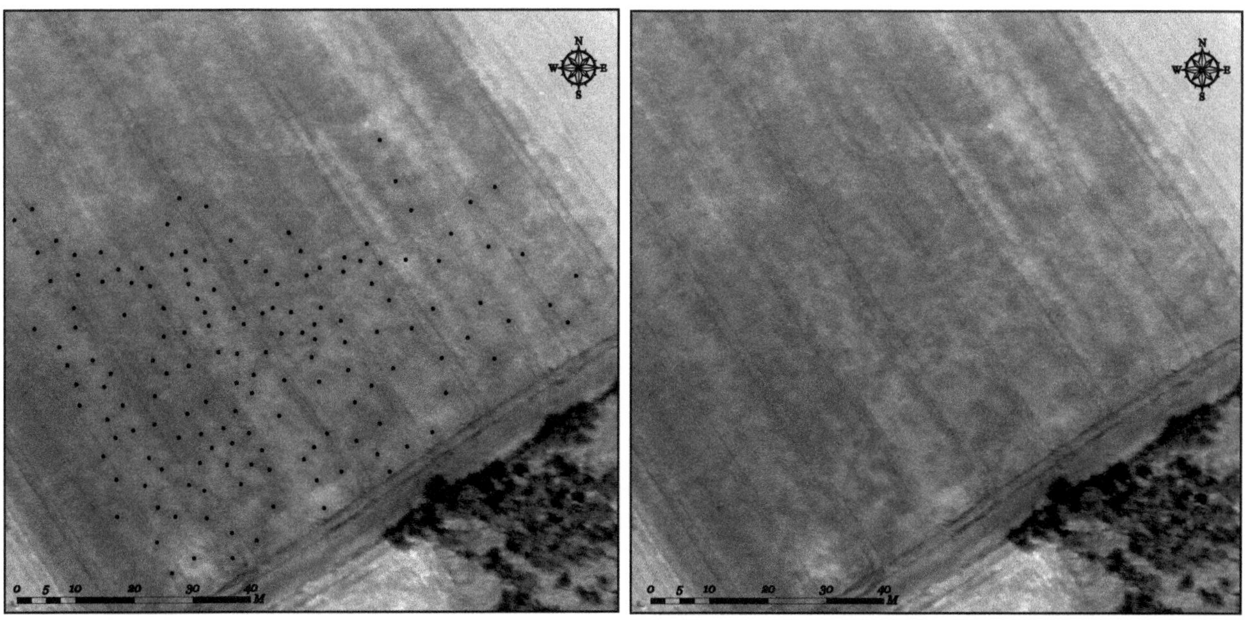

Fig. 2: Location and density of mine shafts interpreted according to aerial photographs in the northern sector of the site (Photography: P. Laforest Service Archéologique Départemental des Yvelines, adjustment: Terra NovA).

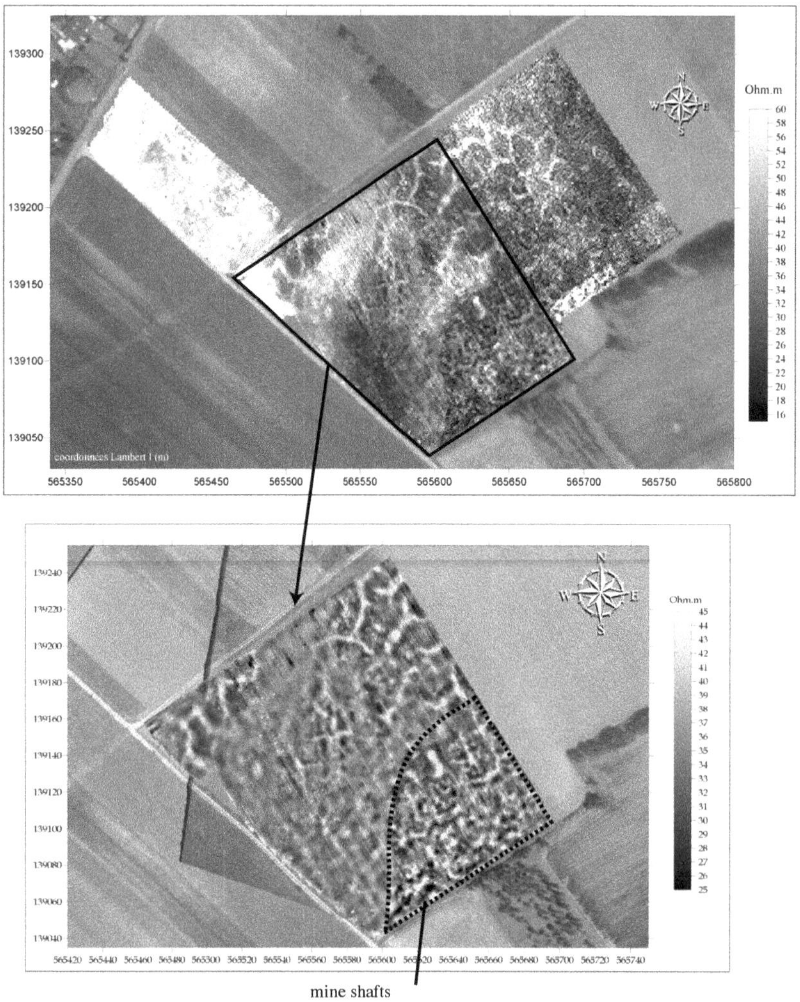

Fig. 4: Electrical resistivity map at 1 m treated with a high-pass filter in the middle part of the site (Terra NovA document, 2003, Riquier C. DAO).

These distances suggest the presence of shallow extraction pits. By comparing these data with the map of the Jablines surface mine, we can envisage pits at a depth of between 2 and 4 m for those which are closest together, as in the North zone at Jablines, where the land is stripped and where structures are the closest together and the most shallow (Bostyn and Lanchon eds 1992). Without prejudging the depth of mining structures at Flins since they have not been excavated, the 2003 experiment identified a layer of flint at an approximate depth of 2.50 m. The convergence of these various observations suggests that some of the extraction sites were dug to a shallow depth.

The spatial extent of the mine could not, however, be determined with reliable precision, as most of the field surfaces photographed revealed little because of a lack of vegetation. Thus, geophysical and pedestrian surveys were conducted concurrently.

2.2. Geophysical surveying

Two geophysical surveying methods were tested on a surface where aerial photography had suggested mine shafts: one using a magnetometer and one using electrical methods (Aubry *et al.* 2002, 2003). After this first experiment, the magnetic method was found to be inappropriate because it could not detect the mine shafts; the perturbations recorded by this method were caused by changes in the soil composition and agricultural activities. Yet, on the basis of a measurement for every metre, the electrical survey confirmed the presence of structures visible in aerial photography, with a good resolution.

Having thus been proven effective, this method was applied to other plots, in particular those for which the aerial photography had revealed nothing. We applied the electrical method to cover a total surface area of more than 5 hectares

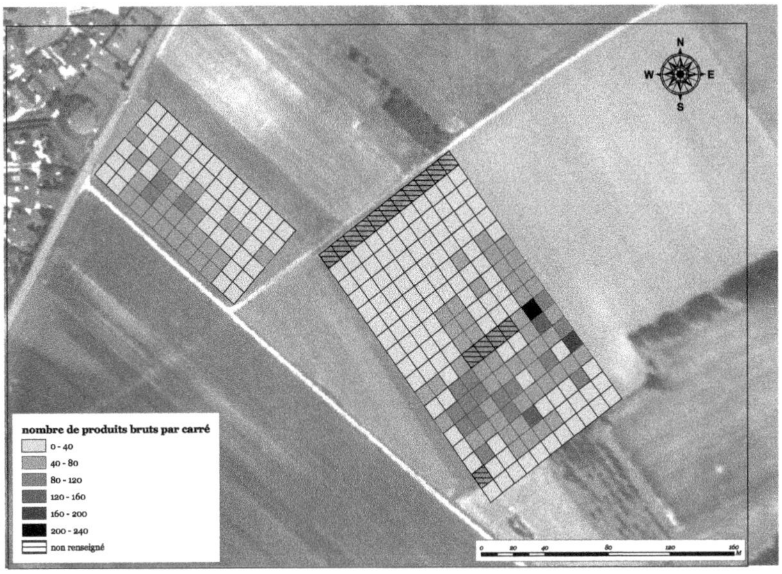

Fig. 5a: Mapping the number of chips per square.

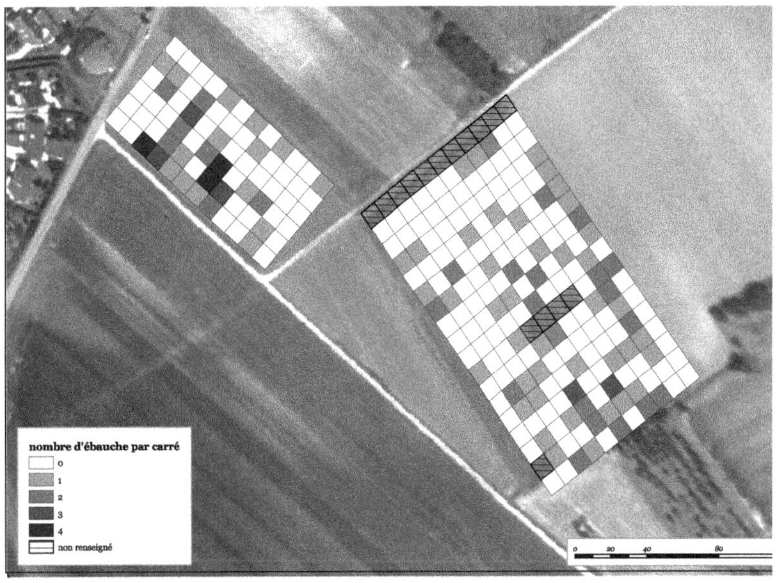

Fig. 5b: Mapping the number of roughouts per square (2002-2003 survey, Riquier C. DAO).

(Fig. 4). The interpretation of the resistivity maps allowed us to target the southernmost point of the research area as the development zone for extraction structures, while the northernmost sectors do not seem to have not been used for underground extraction. However, periglacial phenomena (frost polygons) are clearly visible.

2.3 Pedestrian surveys

Systematic pedestrian surveys were carried out in two successive campaigns on two different plots covering 3 hectares. The Northwest Sector was prospected in 2002, the Southeast Sector in 2003.

The sampling was at a 25% rate for 100 m2 squares. Furthermore, a quick walk over the whole area allowed us to fix the limits of the site approximately. The relative importance of material recovered allowed us to define areas of concentration, corresponding to axe blade knapping activities. The artefacts found were mostly chips, tools of poor craftsmanship and fragments of bifacial pieces. And absence of ceramics and grinding equipment correlated with the small number of tools suggests that domestic activity on the site was uncommon. The area was used as a mine and workshop. A few test boulders and some large sandstone hammerstones were also collected.

The distribution map of the number of chips per square (Fig. 5a), presents a greater concentration in the Southeast Sector, within a strip 50 meters wide, situated northeast / southwest, that follows the level curve of 110 m NGF. This sector has the highest densities (> 120 chips/ square). Two

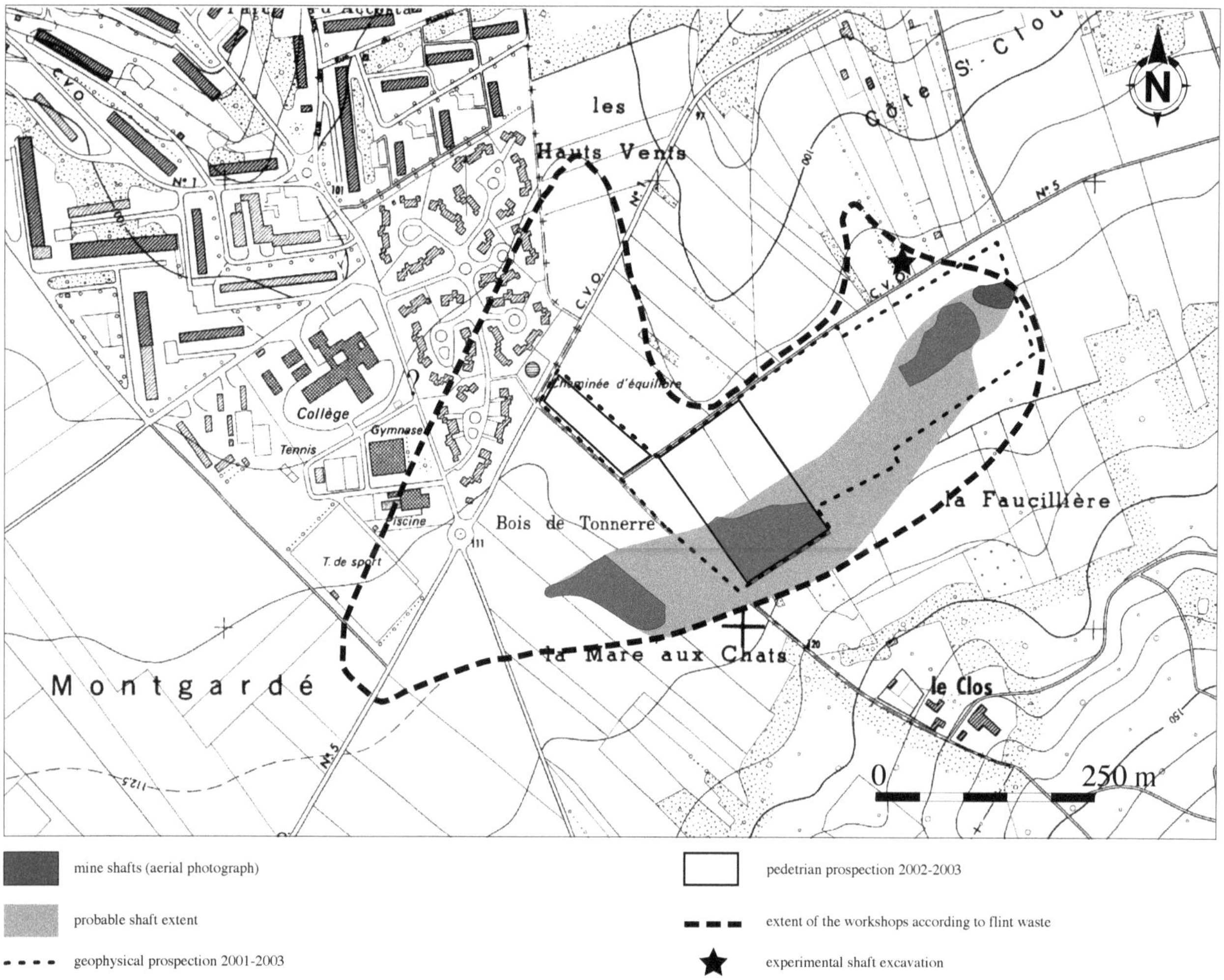

Fig. 6: Extent of the site and summary of the surveys.

areas of lower density appear in this sector at the southeast and northwest boundaries of the plot, at the highest and lowest points. The density is much lower (to 90 chips / square) in the Northwest Sector. An area of concentration emerges in the Northwest, along the path to the site. The distribution map of all rough pieces offers the same image (Fig. 6). However, the distribution map of the roughouts (Fig. 5b) gives a contrasting image, with a higher density in the Northwest Sector. In this sector, the concentration of roughouts coincides with that of the chips, while in the other sector, the distribution is more diffuse across the entirety of the surface and does not outline any concentration different from the chips.

2.4. Conclusion

A comparison between the data offered by the pedestrian surveys, the geophysical data and the aerial photography (Fig. 6) is only possible for the Southeast Sector. In this case, there is a strong correlation between the areas of debitage waste and the anomalies, which are concentrated in the southern half of the plot. This is also, topographically speaking, the highest zone and the one most subject to erosion. In the Northwest Sector, where only pedestrian surveys and geophysical data were possible, no anomaly was detectable by geophysical surveying, while relatively high concentrations of debitage waste, tools and roughouts were confirmed. This sector is thus representative of a zone restricted to knapping, apart from the areas of extraction.

3. PRODUCTION

The analysis of the waste flint and the tools found during surface collection showed that the knapping activities at the mine were primarily related to the production of axes. Very few other tools usually associated with settlement (scrapers, blades) were found. The flint artefacts found at the mine may be divided into several categories representing various stages of the *chaîne opératoire* of axe production, but also of other types of products:

- rough boulders and test boulders
- debitage waste including chips, blades, and splinters
- bifacial fragments in various manufacture stages, or other flint pieces deriving from this *chaîne opératoire*, from the removal platelet down to the retouched axe
- nuclei and other products not connected with the *chaîne opératoire* of axe manufacture
- debris
- flake tools, blades or debris

Since the main purpose for the study of the axe blades is to define the causes of abandonment and then to assess the skill level of the knappers, several criteria were chosen for systematic recording (in addition to the usual criteria of size, state of preservation, etc.): the quality of the raw material, the nature of the blank, and the problems associated with knapping and related technical errors.

The raw material can be classified as good, moderate or poor quality. To define these quality levels, we detailed the imperfections of the raw material: an intrusive cortex, weak silicification, former fracture surface and cavities. A weak blank is defined by a failure to overcome technical imperfections in the raw material and to obtain an axe blade; this corresponds in our study to the criterion of rejection of pieces in the *chaîne opératoire*. A blank of moderate quality may have imperfect silicification and thin slab heterogeneity, but still serve as the basis for making a finished product.

Symmetry in the sections and faces was noted, with a presence or absence of the axis of symmetry on each piece. The edge, where existing, was described, including whether it was of convex, straight or irregular form, its symmetry and its regularisation.

Technical errors were described for each face, with detailed notes regarding hinged flakes, due to the quality of the raw material or the knapping, plunging flakes, concave flakes, overhewn or minor flakes. The edges are straight, irregular, or considered regularized when they have been retouched along the entire length of the edge - or at least 3/4 of the piece - or crushed with a hammerstone.

The angles of the edges are considered irregular when they vary at least 30 ° from the edge. Indeed, some axe blades have angles on part of the edge at more than 90 degrees, a result considered as a technical error.

The bifacial pieces from the surface collections were studied in greater detail. Within a total of 542 pieces, only those found completely intact were retained, reducing the total sample to 428. The blanks are platelets, plate fragments, chips, chips due to frost, or classed as uncertain. The main blanks used are mostly raw plates and chips, in equal proportion (Fig. 7). A 6% ratio of frost chips, identified solely on the steps prior to the knapped axe, should also be noted (Fig. 10). Some of the pieces bear ancient fractures related to the knapping, or to the raw material. However, knappers continued knapping and hewing and the fracturing was not the main cause of rejection.

The main difficulty of our study is to determine what led to completion of the pieces and / or the skill level of the knappers. While it seems obvious for certain criteria, such as errors in size, all other indications are difficult to judge. The presence of cortex and its extent are the criteria generally taken into account to define the degree of progress of the pieces, but these can also testify to the expertise of the knapper, since the presence of marginal chips is a telltale sign of a low level of competence. We must therefore modulate several criteria to determine reasons for abandonment, and to attribute the pieces to a particular technical stage.

	number	%
unifacial preparation	10	2,3%
uni- & bifacial preparation	40	9,3%
Bifacial unilateral preparation	23	5,4%
Bifacial bilateral preparation	142	33,2%
Rough-out	144	33,6%
axe	69	16,1%
Total	428	

Fig. 7: Number of pieces at Flins-sur-Seine showing various stages in the production of axe manufacturing.

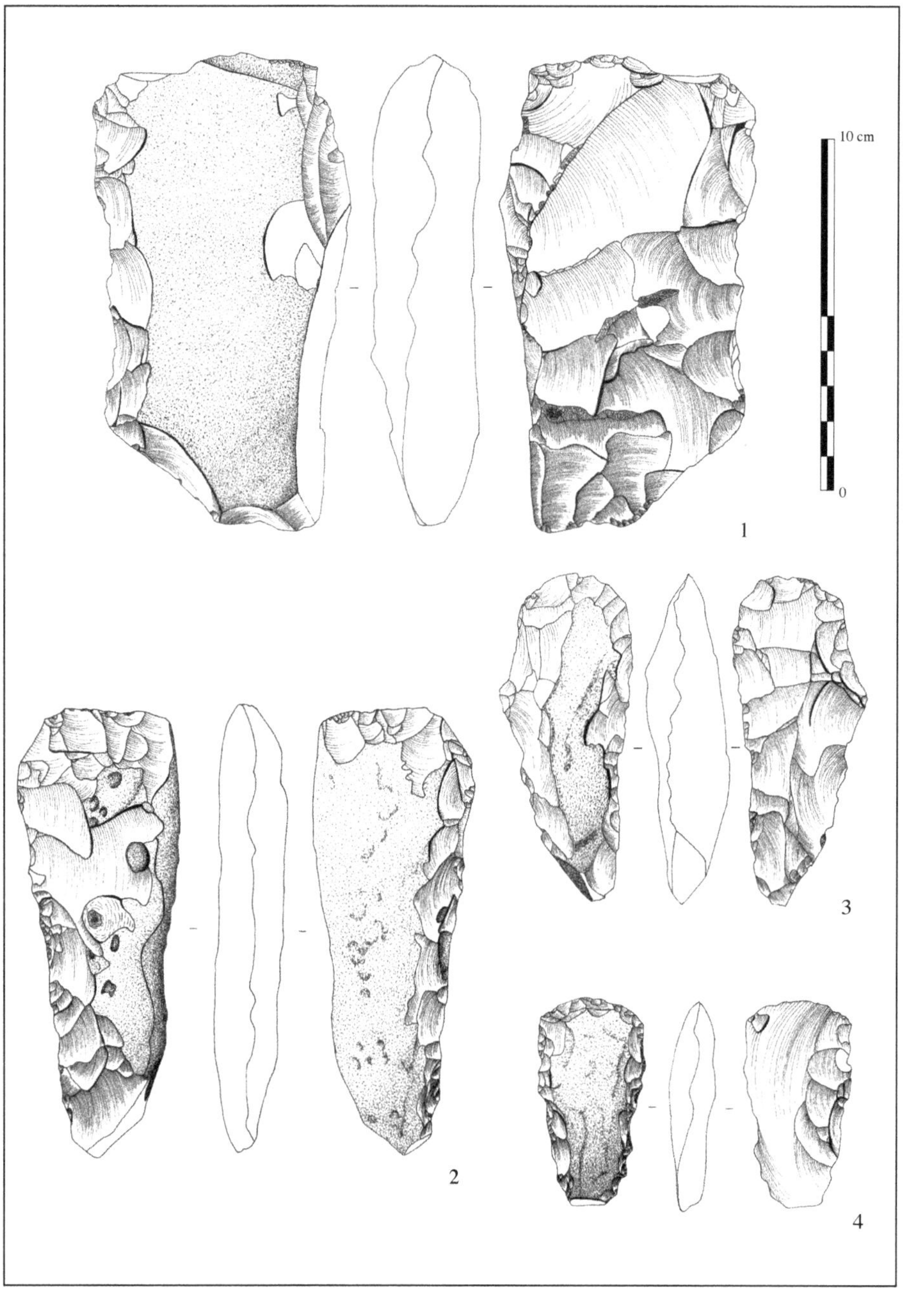

Fig. 8: 1. Bifacial and bilateral preparation and 2. Bifacial unilateral preparation (Drawing Carmine A. Lo). Flins-sur-Seine.

The technological study helped clarify the various stages of the *chaîne opératoire* of production of axe blades. The blank used is mostly of rough slab, or previously struck chips.

Despite technical limitations related to the morphology of the blanks, the *chaîne opératoire* vary only slightly. These sequences are divided into four technical stages: unifacial then bifacial preparation, unilateral and bilateral, the roughing out, regularization/ retouching and polishing. Traces of polishing are not attested at the mine. The bifacial pieces found are representative of the first three stages of the process, from unifacial preparation to the finished product (Fig. 7).

The early stages of shaping occur the least frequently in the sample (17% of the total sample), in particular fragments in unifacial unilateral preparation, or bifacial unilateral preparation. The more advanced stages, i.e., of bifacial and bilateral preparation (Fig. 8) and the waste flakes (Fig. 9) are the most common (66.8% of the total sample). The final step, the axe or preform ready to polish, makes up a little over 16% of the total sample (Fig. 10).

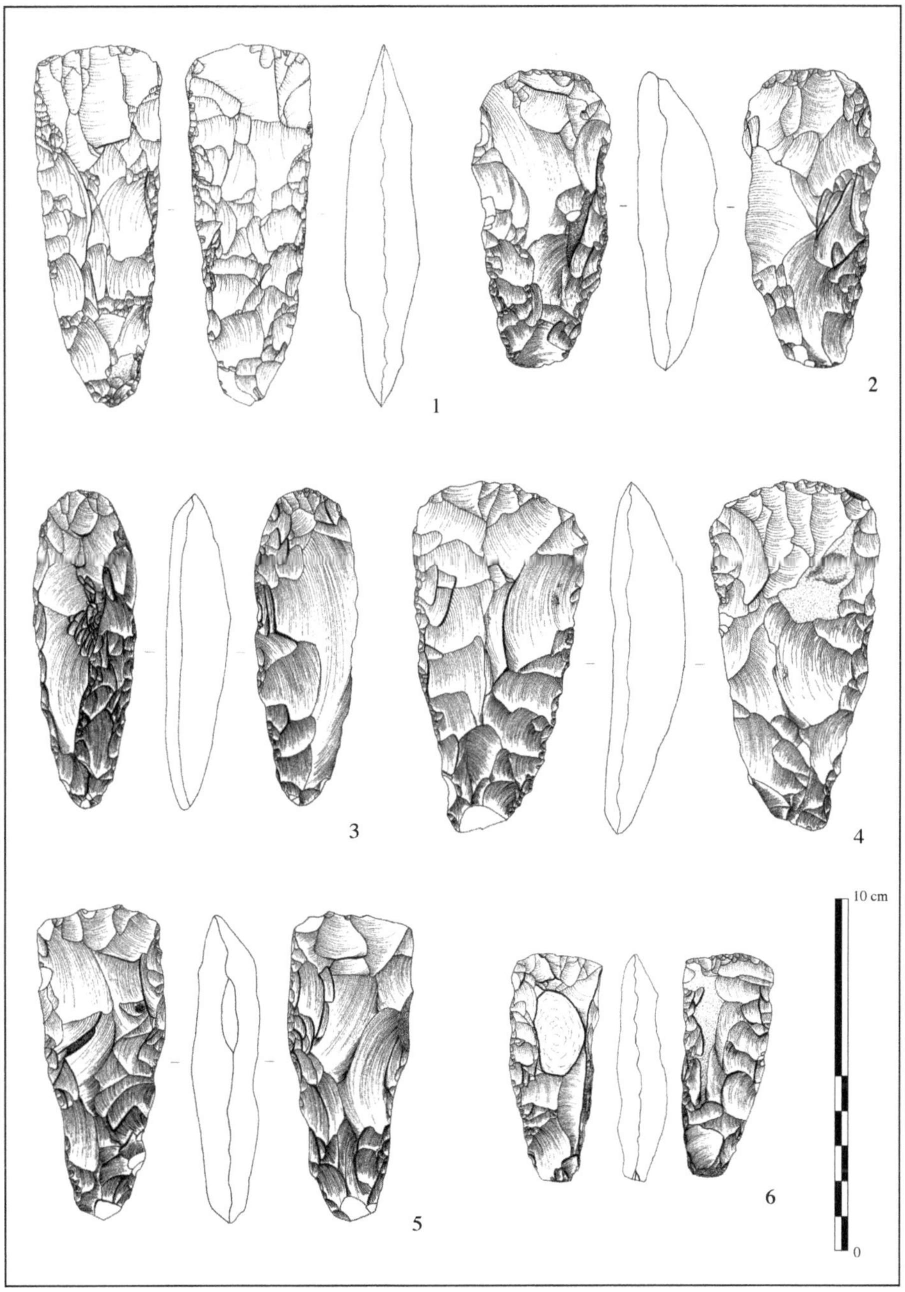

Fig. 9: Roughouts. Flins-sur-Seine (Drawing Carmine A. Lo).

The nodules found are between 8 and 14 cm, though larger pieces exist as well. Their lengths are between 15 and 25 cm. The pieces obtained from levelling up the stone facing are larger, especially when the blank is a small plate, as the rough-hewn large flakes broken off with a hammerstone greatly reduce the size of the original blank (Fig. 11). The pieces have an average length of 12.5 cm for those of bifacial preparation, 10.5 cm for axes, with a low coefficient of variation. These two categories make up the largest and the smallest pieces (from 6.3 to 26.5 cm). The average widths are 5.5 cm for bifacial preparations, 4.2 cm for the axes; with thicknesses of 2.9 cm for the bifacial preparations, 2.5 cm for the axes.

These flint pieces were discarded mostly because of the poor quality of the raw material, which contains cavities (32%), intrusive cortex (33%) or a weak silicification (15%). Only 46% of the bifacial and bilateral preparations are made of good quality blanks, with 53% for the axes. However, if one takes into account three other criteria, there arises a variety of problems that led to discarding these pieces. Some of them are too hollow on one or both sides: 34%

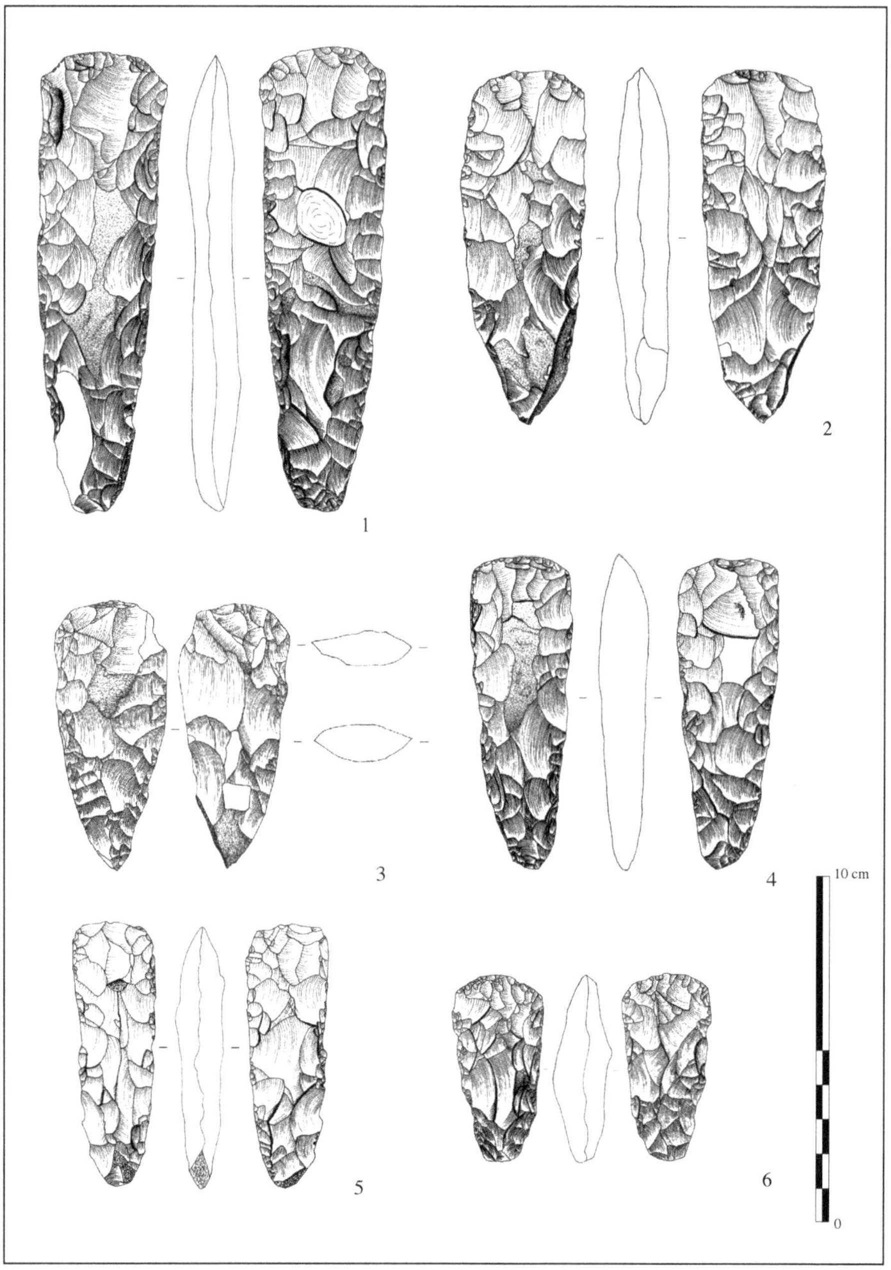

Fig. 10: Axes. Flins-sur-Seine (Drawing Carmine A. Lo).

of bifacial and bilateral preparations, 37% of the roughouts but only 12% for the axes. Another error recognized on discarded fragments is repeated blows, relentlessness (*acharnement*) on one or both edges. Sometimes percussion action along the edge caused it to splinter irregularly, creating a defect impossible to correct, as was seen in 61% of the bifacial preparations, 55% of the roughouts and 40% of the axes. Manual awkwardness is without doubt one of the most frequently observed problems, impugning the poor level of competence of the knapper. To conclude this summary review of inhibitory over-levelling, we will refer to the statistics: although less frequent, over-levelling was revealed on 19% of the bifacial bilateral preparations, on 13% of the roughouts and on 9% of the axes.

An overall analysis of the products found at the mine suggests that the general level of quality was moderate, both in regard to the rough blanks as to the levels of expertise. These results contrast with the level of investment required to acquire the flint, and also with the quality of the products found outside the context of the mine, a point to which we shall return later. How should we interpret

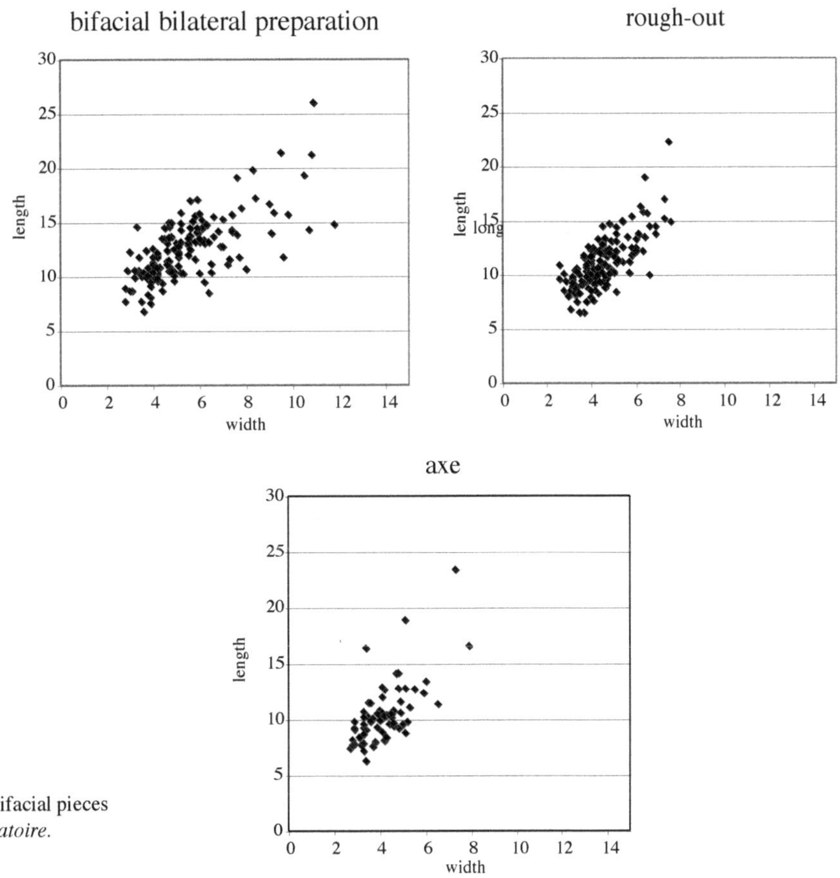

Fig. 11: Length width relationship of bifacial pieces according to stages in the *chaîne opératoire*.

these findings from a socio-economic viewpoint? It seems that we are faced with two types of productions, with varying end results. A production of large axe blades (30 cm), probably made from platelets selected from the most homogeneous rock, was carried out at the mine by probably experienced knappers. These products were intended to circulate at shorter or longer distances and are only rarely found among the extractive waste. This consideration is logical, since we can expect a very low failure rate on the part of experienced knappers. With this in mind, we can consider the presence of apprentice knappers at the location of the extraction, where the raw material is abundant, thus allowing for a certain amount of waste. A poor mastery of the technical and visualizing know-how (Pelegrin 1991) would result in failed production at different stages of the *chaîne opératoire*. One can also consider that the apprentice knappers were intended to produce pieces of smaller size, the manufacture of which is less risky. The tools produced may have been used on the spot or taken to nearby habitats.

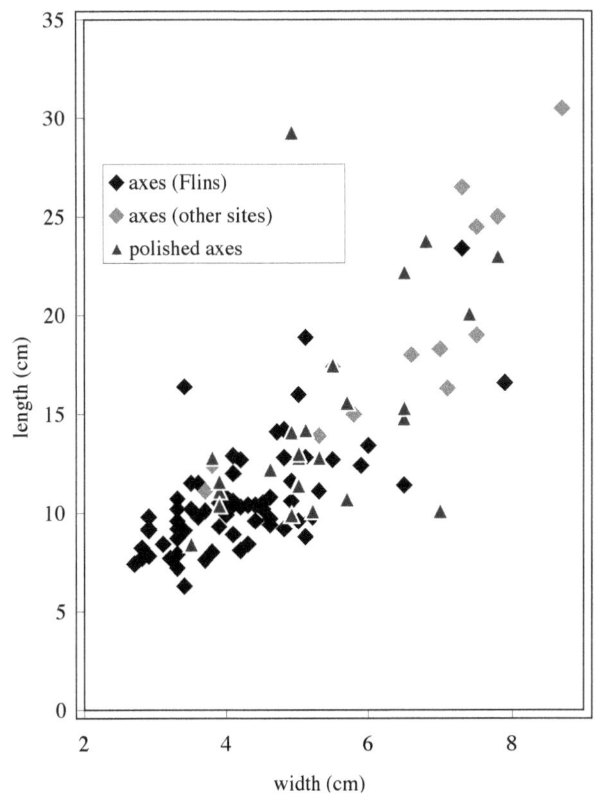

Fig. 12: Comparison of the scale of knapped axes from the mine, with knapped and polished axes at a regional level.

4. PRODUCT DIFFUSION

To assess the extent of product diffusion at the regional level, cartographic work was carried out, mapping the distribution of polished, percussioned and unfinished axe blades. This mapping focused primarily on the Departments of Yvelines and Val d'Oise, but also extended further to the west and southwest.

The flint pieces found outside the mine at a regional level are sometimes larger, 15 cm to 30 cm, an observation which applies to both axes and polished axes (Fig. 12). The status of these large artefacts, unfortunately found out of context, such as in Bonnières-sur-Seine, raises the question of their function (Fig. 13, n° 1). They do not seem to have been used (the sharp edge was retained), and could be classified in the category of objects of prestige. In the context of settlements, the flint artefacts are smaller, often recycled, with a reknapped edge; some were used as hammerstones, and highly fragmented (Fig. 13, n° 2).

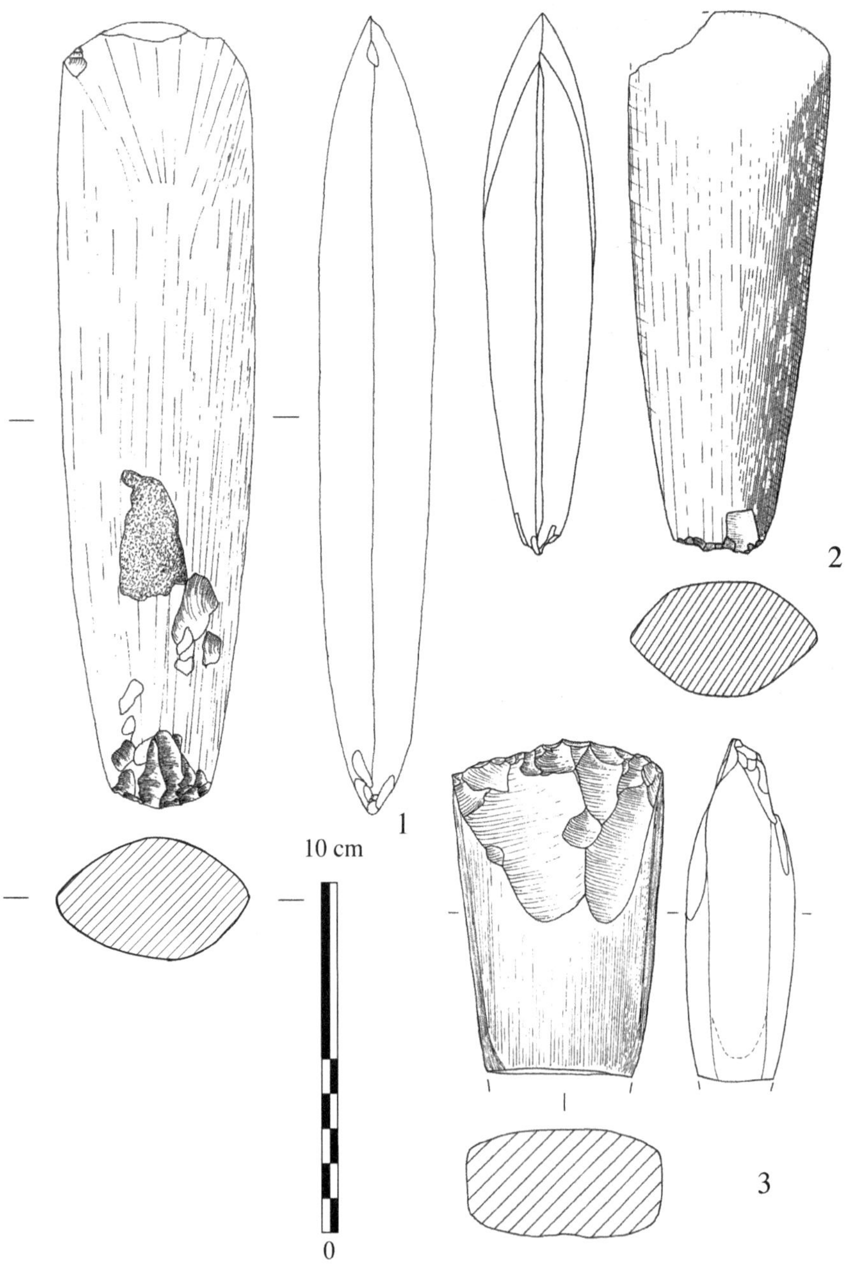

Fig. 13: Polished axes. 1. Discovered out of context. 2. Late Neolithic collective grave Val de Reuil 'Les Varennes', 3. Early Neolithic settlement Neauphle-le-Vieux (1: drawing Lo Carmine A.; 2: Billard C. 1995, fig. 13; 3: drawing Martial E.).

Distribution maps of the products show differences depending on the operating stages. The early stages of preparation, in particular bilateral and bifacial preparations, are found almost exclusively at the mining and surrounding areas (Fig. 14a). The few points further afield, however, show that some unfinished pieces were circulated in small quantities.

Away from the mine, but probably in direct correlation to it, the highest concentrations correspond to probable flint knapping workshop sites (Fig. 15). It is useful to recall that several grooved abraders have been found in the immediate area close to another workshop at Jumeauville. During the Neolithic period, the banks of the Mauldre must have been a particularly dynamic area of activity focusing around the working of flint (Fig. 15). One central site could have played a role in an interregional exchange system.

The distribution of roughouts and axes is more scattered, showing a diffusion to the North, on the right bank of the Seine, and about 60 km to the Southwest (Fig. 14b and 16 a). Polished axes were circulated at a much wider distance and a much larger scale (Fig. 16b).

There is still a difference in quantity between the regions south of the Seine and those in the north. The quantity of polished axes in Bartonian flint is particularly low in the Val d'Oise (16%): in this case, the Seine must have been a natural obstacle (Fig. 17). The presence of workshops to make secondary flint axes in the two regions can also provide some explanation. In the Yvelines, the two raw materials are represented in the same way, so they must have been in direct competition.

To conclude, it should be noted that some of the flint artefacts travelled at even greater distances, towards Normandy, in the Seine valley and more than 150 km away into the plains of Caen and Argentan (Watté 2007, Bostyn *et al.* in press b).

5. CHRONO-CULTURAL CONTEXT

In the absence of direct excavation-derived data at the Flins mine, its chronology must be estimated from other excavated sites. Unfortunately, the longest, most complete products of this mine were found out of context (Fig. 13, n°1). Polished axes in Bartonian flint are very rare at the sites of the former Danubian Neolithic (Villeneuve-Saint-Germain). For this time frame, polished pieces from the Paris Basin are equally rare, and usually made of material other than flint, such as quartzite sandstone or resistant exogenous rock. However, it should be noted that a fragment of a polished axe in Bartonian flint was found at the site of Neauphle, Yvelines (Giligny *et al.* 1996 and Fig. 13, n°3), and an incomplete bifacial piece was found at the site of Ocquerre, Seine-et-Marne (Praud *et al.* 2002).

Bartonian flint axes become common during the Middle Neolithic. The Neolithic settlement site at Louviers (Eure), fifty kilometres west of the mine, illustrates the circulation of finished products from this mineral source at that time. At Louviers, tertiary flint makes up 1/3 of all axe blades on the site (Giligny ed. 2005).

The deposition of axes in some Neolithic graves indicates that the production lasted until the late Neolithic, with a wide chronological bracket concerning the use and reuse of collective graves over a long period of time (Seine-Oise - Campaniforme Marne). These collective graves are found in Luzarches Compans 'Le Val d'Oise, Presles II' La Pierre Plate, 'L'Etang La Ville' Cher Arpent (Yam 2005), and Val de Reuil 'Les Varennes' (Billard *et al.* 1995 : Fig. 13, n° 2).

CONCLUSION

These initial analyses show a relatively complex organisation in the production of axe blades, in their socio-economic organisation, where the products seem to have been made by knappers with highly variable levels of skill, and also in the spatial distribution between the areas of extraction strictly speaking the workshop areas, and the polishing zones. Continuous evidence of tertiary flint axes in both excavation and surface campaigns is a prerequisite for a study of their distribution. Such research will better measure the importance of the mine of Flins at a regional level, which extends more widely into the western Paris Basin, but also will help evaluate the relative positions of the two mines identified with certainty in the Tertiary Basin, that of Flins and that of Jablines.

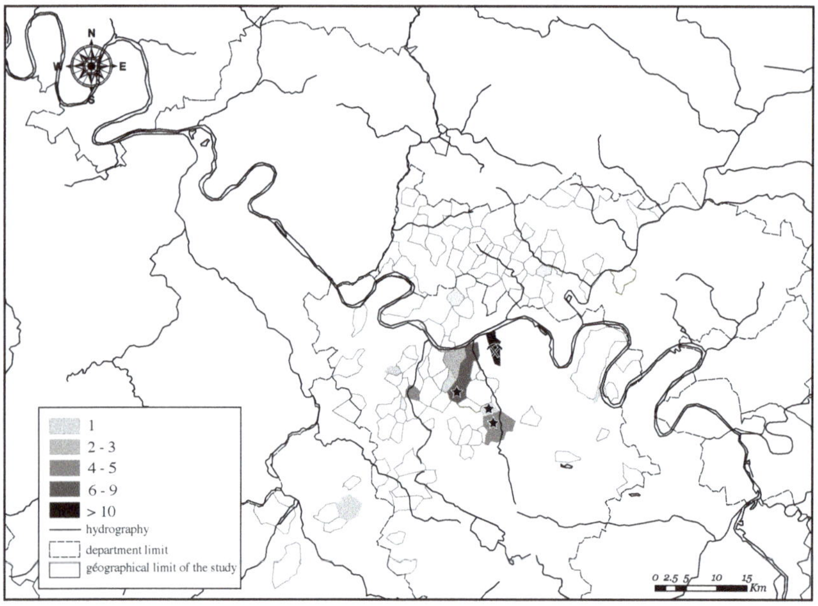

a

Fig. 14a: Distribution map. Bifacial preparations.

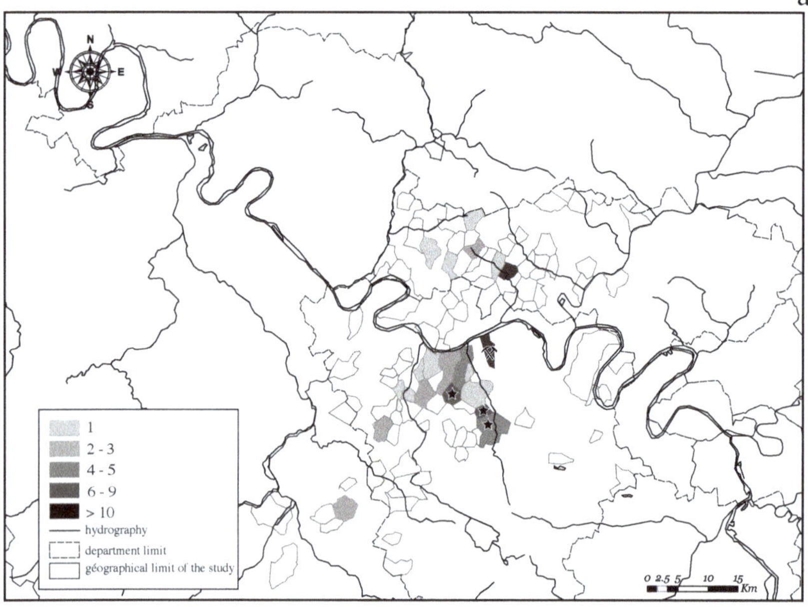

b

Fig. 14b: Distribution map. Roughouts.

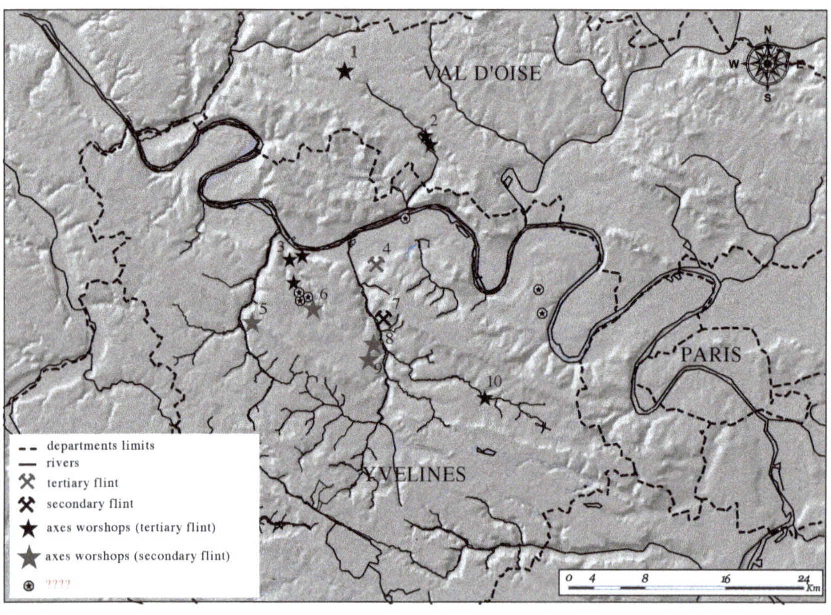

Fig. 15: Map of the different workshop and polishing sites along the Mauldre river.

1- Banthelu
2- Longuesse
3- Guerville
4- Flins
5- Rosay
6- Jumeauville
7- Maule
8- Montainville
9- Beynes
10- Villepreux

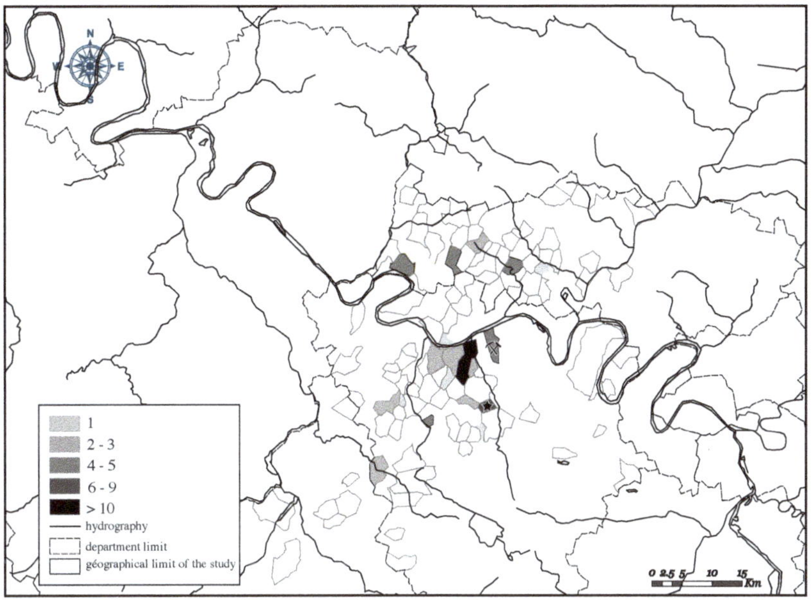

Fig. 16a: Distribution map. Axes.

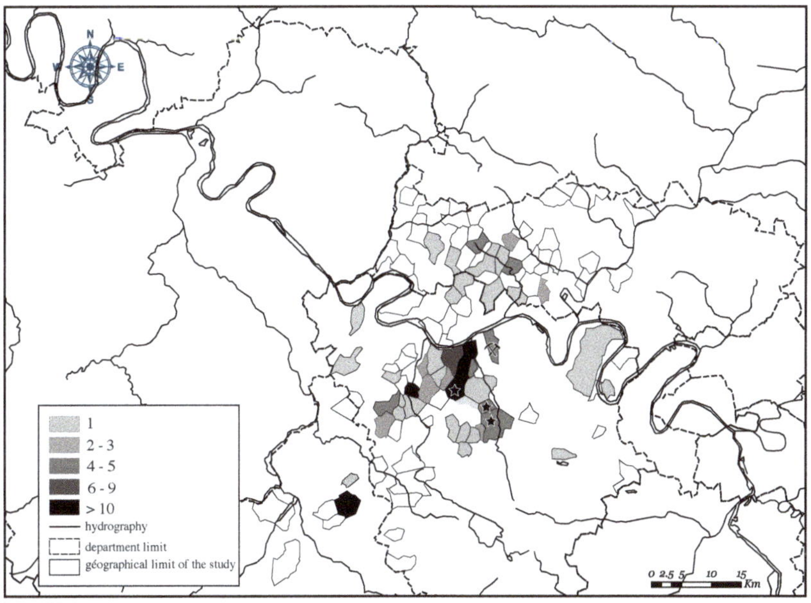

Fig. 16b: Distribution map. Polished axes.

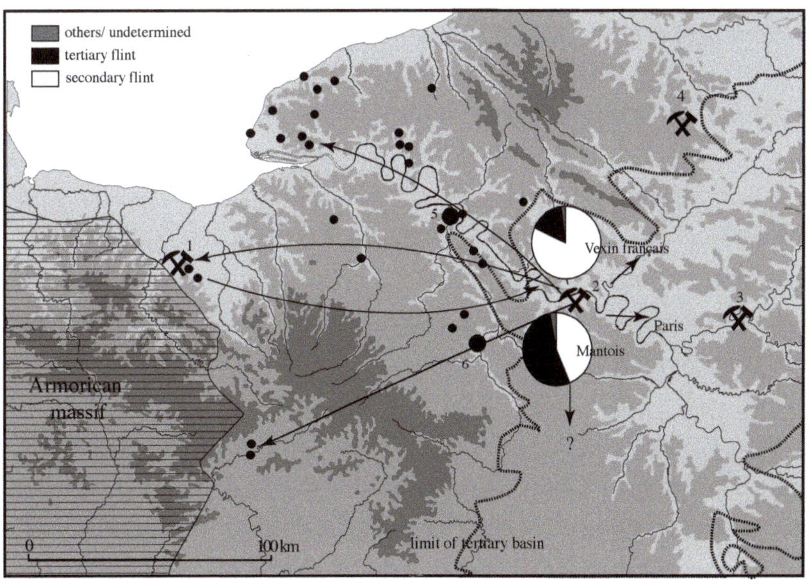

mining complex • sites with bartonian axes ● bartonian axes >1

1 : Bretteville-le-Rabet, 2. Flins-sur-Seine, 3. Jablines, 4. Hardivilliers-Troussencourt, 5. Louviers, 6. Fort-Harrouard

Fig. 17: Nature of material used for polished axes in the north and south of the Seine river.

Françoise Bostyn
Inrap - UMR 7041
36 allée Thalès
59650 Villeneuve d'ascq
e-mail: francoise.bostyn@inrap.fr

François Giligny
Université de Paris I - UMR 7041
MAE 21, avenue de l'université
92023 Nanterre
e-mail: francois.giligny@univ-paris1.fr

Adrienne Lo Carmine
Université de Paris I - UMR 7041
MAE 21, avenue de l'université
92023 Nanterre
e-mail: adrienne_locarmine@yahoo.fr

REFERENCES

Aubry, L. and Barrés, E. 2002. *Prospections géophysiques, Minières néolithiques de Flins-sur-Seine. Campagne d'acquisition.*

Aubry, L., Barrès, E. and Marmet, E. 2003. *Prospections géophysiques, Minières néolithiques de Flins-sur-Seine. Synthèse des campagnes 2002-2003*, Terra NovA.

Billard, C., Guillon, M., Sunder, F. and Arbogast, R-M. 1995. La sépulture collective du Val-de-Rueil 'Les Varennes' et les structures associées de l'Age du Bronze. In *Actes du Colloque interrégional sur le Néolithique, Evreux 1993*, Revue Archéologique de l'Ouest, supplement 7, 155-182.

Billard, C., Guillon, M. and Verron, G. (eds) with Argobast, R.-M., Cedran, Y., Sunder, F., Dasturgue, J., Gervais, V., Leuge, F., Querre, G. and Salanova, L. *Un système de sépultures collectives au Néolithique final à Val de Reuil et Portejoie (Eure),* unpublished.

Bostyn, F. 2003. De la lame à la hache : contextes géologiques et socio-économiques des productions en silex Tertiaire bartonien du Bassin parisien au Néolithique. In *Les matières premières lithiques en préhistoire, actes de la table ronde d'Aurillac, 20-22 juin 2002,* Préhistoire du Sud-Ouest, supplément 5, 63-70.

Bostyn, F. and Lanchon, Y. (eds) with Boguszewski, A., Frugier, C., Jérémie S. *et al.* 1992. *Jablines, Le Haut Château (Seine-et-Marne) : une minière de silex au Néolithique*, Documents d'Archéologie Française 35. Paris : Maison des Sciences de l'Homme.

Bostyn, F., Cayol, N., Giligny, F. and Lo Carmine, A. in press a. Creusement expérimental d'un puits d'extraction de silex sur la minière de Flins-sur-Seine (Yvelines), *Actes du XXe Colloque interrégional sur le Néolithique.* Luxembourg : Archéologie Luxembourgeoise.

Bostyn, F., Couderc, J., Giligny, F., Lethrosne, H., Le Maux, N. and Lo Carmine, A. in press b. Ateliers de fabrication de haches et minières à silex dans l'Ouest de l'Île–de-France (Yvelines, Val d'Oise). Table ronde de la Société Préhistorique Française, mars 2007.

Bourgeois, L. 1982. *Inventaire d'une série lithique provenant de Flins-sur-Seine (Yvelines). Rapport.* Versailles: Service Archéologique Départemental des Yvelines.

Bricon, D. and Vidal, G. 1983. Haches taillées de grandes dimensions de la région de Flins. *Bulletin du Centre de Recherches Archéologiques de la Région Mantaise* 6, 19-20.

Daniel, R. 1965. Les ateliers campigniens de Flins (Seine-et-Marne). *Bulletin de la Société Préhistorique Française* 62, 279-286.

Giligny, F., Martial, E. and Praud, I. 1996. Le Villeneuve-Saint-Germain de Neauphle-le-Vieux 'le Moulin de Lettrée' (Yvelines), *Internéo* 1, 15-32.

Giligny, F., Martial, E. and Praud, I., with Le Gall, J. 1998a. *L'occupation des Yvelines au Néolithique ancien: premiers éléments. Rapport d'étude.* Versailles: Service Archéologique Départemental des Yvelines.

Giligny, F., Martial, E. and Praud, I., with Bostyn, F., Le Gall, J., 1998b. Premiers éléments sur l'occupation des Yvelines au Néolithique. *Internéo* 2, 43-56.

Giligny, F. and Lo Carmine, A. 2001. *Le Néolithique des Yvelines. Rapport d'étude 2001*, UMR 7041 – Nanterre.

Versailles : Service Archéologique Départemental des Yvelines.

Giligny, F. (ed.), Bostyn, F., Couderc, J., Durand, S., Durand, J., Lo Carmine, A., Martial, Praud, I. and Riquier, C. 2005. *Le Néolithique des Yvelines. Rapport intermédiaire du Projet Collectif de Recherches.*

Giligny, F. (ed) 2005 - Louviers "la Vilette" (Eure) : un site Néolithique moyen en zone humide. *Documents Archéologiques de l'Ouest*, 2005, 343 pages.

Lo Carmine, A. 2000. *La production de haches en silex tertiaire dans le Nord des Yvelines*. Mémoire de maîtrise de l'Université de Paris 1.

Lo Carmine, A. 2002. *Grandes lames de haches et ateliers de fabrication dans le nord des Yvelines*. Mémoire de DEA de l'Université de Paris 1.

Pelegrin, J. 1991. Les savoir-faire: une très longue histoire. *Terrain (Carnets du Patrimoine Ethnologique)* 16, 106-113.

Pelegrin, J. 1995. Réflexions méthodologiques sur l'étude de séries lithiques en contexte d'atelier ou de mine. In J. Pelegrin and A. Richard (eds), *Les mines de silex au Néolithique en Europe,* Documents Préhistoriques, Edition du C.T.H.S., 7, 159-166.

Praud, I., Bostyn, F., Martial, E. and Michel, L. 2002. Un site Villeneuve-Saint-Germain dans la vallée de l'Ourcq. *Internéo* 4, 13-22.

Silvestre de Sacy, L. and Baudouin, M. 1926. La station campignienne du Clos, à Flins-sur-Seine (Seine-et-Oise). *Bulletin de la Société des Sciences de Seine-et-Oise* fasc. 4, 2e série, t. VII, 49-62.

Silvestre de Sacy, L. and Baudouin, M. 1927. La station campignienne du Clos, à Flins-sur-Seine (Seine-et-Oise). *Bulletin de la Société des Sciences de Seine-et-Oise* fasc. 4, 2e série, t. VIII, 50-58.

Watté, J.-P. 2007. Objets en silex tertiaire importés en Haute-Normandie. *Haute-Normandie archéologique* 12, 53-67.

Yam, S. 2005. *Le mobilier funéraire des allées couvertes du Val d'Oise.* Mémoire de Maîtrise de l'Université de Paris 1, 2 vol.

THE KRZEMIONKI FLINT MINES LATEST UNDERGROUND RESEARCH 2001-2004

Jerzy Bąbel

Abstract: In 2001-2004 the new underground gallery was made in the Neolithic flint mine at Krzemionki Opatowskie, southern Poland. The gallery passes through the area of eight shafts. The author with the team from Archaeological and Historical Museum at Ostrowiec Świętokrzyski carried on the archeological invesigations during the mining works connected with the preparations of the gallery. The new observations were made concerning the stratigraphy of rubble tips, traces of works, tools, illumination and ventilation in the prehistorical shafts and so on. The new 10 radiocarbon datings from pices of charcoal were obtained. Now the udergound touristic route at Krzemionki is approxiately 460 m in length.

Keywords: Mining field, Neolithic flint mines, flint workshops, fireplaces in mines, ventilation, charcoal drawings, wooden torch, dry filling, wooden, stone and antler tools, wooden constructions, traces of works, traces of rainwater in mines, underground tourist tail.

In the Mesozoic margin of the Świętokrzyskie Mountains (=Holy Cross Mts.) in southern Poland, there are vast outcrops of various kinds of flint deposits, and many prehistoric flint mines. Mines of flint showing the characteristic striped texture have been found at Korycizna, Borownia and Ruda Kościelna, along the Kamienna river valley, north of the Holy Cross Mountains, NE of Ostrowiec Świętokrzyski, and also at Krzemionki, 8 km NE of Ostrowiec Świętokrzyski (Budziszewski, Michniak 1983/89). The complex of flint mines at Krzemionki, (also known as Krzemionki Opatowskie, in terms of area of the mining field), is one of the largest in Europe (see: Barber *et al.* 1999, p.58, fig. 5.5). The perfectly preserved 'prehistoric' landscape and underground structure of the Krzemionki mine are of extraordinary importance, both for archaeological investigations and for educational tourism.

The Krzemionki mine was discovered by the geologist Jan Samsonowicz in 1922. Since then, it has been investigated by many Polish researchers. Before the Second World War, the excavations were directed by Zygmunt Szmit (1923, 1927), Józef Żurowski (1925-1927), Stefan Krukowski (1923, 1928-1937). After the war, excavations were continued by Michał Drewko (1945, 1948), Tadeusz Żurowski (1953, 1958-1961), Jan Kowalczyk, Bogdan Balcer and Zygmunt Krzak (1969-1970), Jerzy Bąbel (1979-1984, 2001-2006), Sławomir Sałaciński, Marek Zalewski, Witold Migal (1985-1988) and Wojciech Borkowski (1989-2000), along with many colleagues.

The mining field at Krzemionki is located in an area of exposed Jurassic (Upper Oxfordian) limestone, along the sides of a small syncline (Fig. 1). The parabola-shaped field mine is about 4.5 km long, covering an area of circa

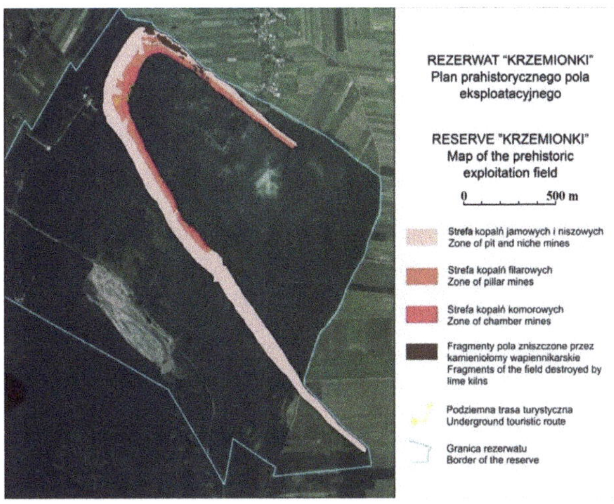

Fig. 1: 'Krzemionki Opatowskie' Reserve, Kielce voivodships. Plan of the prehistoric mining field.

785 000 m^2. The flint concretions occur in three levels. They were exploited to a depth of up to about 9 m, with the deepest shafts dug near the hinge area of the syncline.

The number of mining units is estimated at over 3,500. The shafts were set up 5 to 30 m apart; their depths and shapes depend on the local geological conditions of the flint-bearing limestone level. Ball-shaped and flattened flint concretions were extracted in several ways, depending mainly on the depth of their occurrence in the limestone massif. They were excavated in shallow cavities, in niche mines (2 m deep by 4-5 m wide), in pillar-chamber mines (c. 4.5 m deep), and in chamber mines, up to 8-9 m deep, covering an area of c. 400 m^2 or more.

The mines were exploited from c. 3900 to 1600 BC (according to radiocarbon dating), by people who left artefacts categorized by archaeologists as belonging to the Funnel Beaker Culture (TRB), the Globular Amphorae Culture, and the Mierzanowice Culture. Mesolithic hunters may have known these deposits and used the striped flint even earlier.

A growing Neolithic population, as well as the introduction of the slash-and-burn type of farming, were vital factors that led to the development of flint mining in the Świętokrzyskie region (Holy Cross). Axes made of flint, mostly used for tree-cutting, to clear land as well as to cut wood, were distributed over a range of 250 km around the mines at the time of the development of the Funnel Beaker Culture (c. 3900-2900 BC). However, most shafts at Krzemionki were made by miners of the Globular Amphorae Culture (2900-2500 BC).

The axes then produced, probably for special purposes, have been found as far away as 600 km from the mine. In the Early Bronze Age (the Mierzanowice culture, c. 2200-1600 BC), tools and weapons (axes and arrow-heads) made of flint were distributed in a smaller range of c. 85 km (Balcer and Kowalski 1978, Bąbel J. 1980, Borkowski and Budziszewski 1995).

During the Neolithic Period, the development of more complex flint mining technology resulted in the growth of specialization: at this time professional flint miners emerged. A mine crew consisted of five to ten persons.

Flint was mined both in the warm season (mostly in shallow cavities) and in the cold season (in deep chamber mines).

Sheds were built over the chamber shafts to protect the miners and the mine from rain and snow. The miners used sets of tools made from pieces of flint, from other rocks, and from deer antlers. Those served as wedges, mallets, levers, hoes and pickaxes. There was also an ingenious system to carry the flint output up to the surface. The miners worked underground in a contracted position: half-lying, crouching or kneeling. In order to save work, the excavated tunnels and chambers were only 55-110 cm high. Loosened limestone rubble was disposed of either on the surface, where it was stored in characteristic heaps surrounding the shafts, or it was used to backfill the explored and abandoned chambers. To prevent mine roofs from collapsing, pillars of solid rock were left (in pillar-chamber mines), or supports were made of limestone slabs and rubble. Fires made in the shafts and near their entrances provided air circulation in the mine. The mine was lit by burning resinous chips (Borkowski 1995a and 1995b, Lech 2004).

The flint material obtained was sorted directly underground: only the best quality flint concretions and their fragments were transported up to the surface. Outside the shaft entrance, they were sorted once more, before they were subjected to preliminary shaping. The concretions were broken on a stone anvil and worked with shaping tools made of stone, flint, bone and hard wood. Large amounts of flint waste and aborted semi-products of axes and other tools were abandoned near the shaft entrance (in the site flint workshops). Selected semi-products, or roughly shaped lumps, were taken for further processing to workshops located at a distance of many km from the mine, in the basin of the Kamienna river (e.g. Ćmielów – Gawroniec), where, for instance, axes were polished and completely finished (Balcer 1995 and 2002). Apart from temporary camps built by the miners, there was no permanent settlement in the Krzemionki mining area, because potable water was lacking. Sometimes miners used rainwater, which gathered in some deeper karst hollows, about 250-350 m south of the mining field.

After the prehistoric miners abandoned the flint deposits, the area remained hidden in ancient forest until modern agriculture transformed the landscape, at the beginning of the 20th century, when the village of Krzemionki was created nearby. The inhabitants and the lime-producers partially destroyed the ancient flint mines (among other things, the 'Great Chambers' in the present-day tourist route), in order to gain limestone to produce lime and to use as a fluxing agent for the Ostrowiec steelworks. This destructive

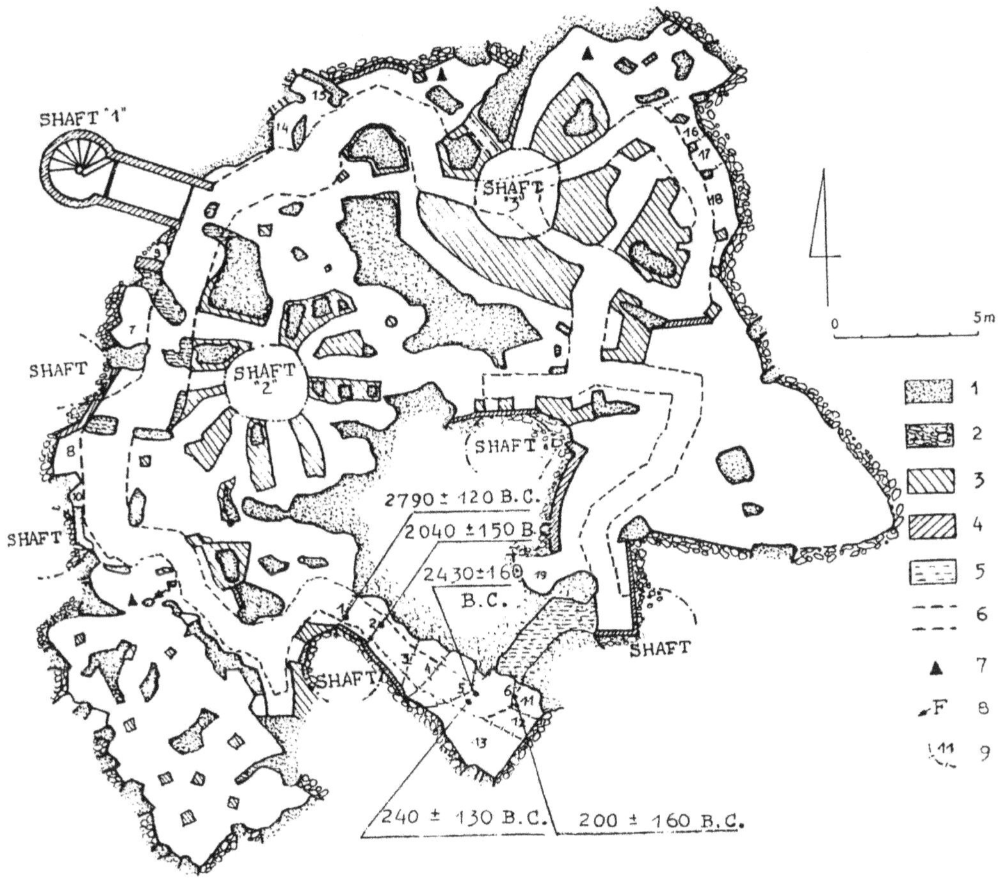

Fig. 2: 'Krzemionki Opatowskie' Reserve, Kielce voivodships. So-called 'Trail no. 1' (acc. to J. Bąbel 1990, fig. 1). Plan of shafts 1, 2, 3 and of locations dated with C-14. 1- solid limestone, 2- dry flling (limestone rubble), 3, 4 - modern protecting devices, 5- solid limestone removed during the working of tourist passage, 6- tunnel route, 7- figures of prehistoric miners, 8- prehistoric drawing of a human figure ('Woman giving birth to a child'), 9- subterranean areas excavated in 1983 and 1984.

limestone exploitation was stopped when the archaeological reserve was established. Its organization began in 1926 (Bąbel 1975). Now, the whole mining field constitutes a Historical Monument of Poland, and a protected Archaeological and Natural Reserve 'Krzemionki Opatowskie' (Hadamik 2006).

One of the important questions was how to show the prehistoric flint mines to contemporary tourists.

Thirty years ago, in 1976, I submitted to public opinion the idea of displaying the Krzemionki mines by excavating underground galleries across the site (Bąbel 1983). I had been inspired by the work carried out since 1964 in the prehistoric mines at Rijckholt near Maastricht in Holland (Engelen F.H.G., 1980; Felder *et al.* 1998). This proposal was debated by specialists for several years, before approval for its construction was granted by the Ministry of Culture and Art.

The project was carried out by an excavation team from the State Archaeological Museum in Warsaw, which I directed during the period from 1979 to 1984. Our work was aided by seism-electrical and Surface Interface Radar (SIR) surveys along the line crossing the main flint-bearing field (Borkowski 1990, Herbich 1993). At that time, the so-called 'Zenon' shaft was sunk and the edge of one of the chambers of mine no. 7/610 was identified. A parallel underground gallery was dug in the region of shafts 1, 2 and 3 (explored in the 1950s). The opening of 'Trail 1' took place in June 1985 (Fig. 2, Bąbel 1986). 'Trail 2' was built around mine no. 7/610 and opened to the public in 1990.

With the growth of tourism in the region around Krzemionki, the idea of joining the trails was once again discussed. The Permanent Krzemionki Reserve Commission (under the auspices of the General Conservation Office, Warsaw) which operated in 1999-2002, under the direction of Professor Jacek Lech, produced a programme for the protection and management of the reserve. One of the points in its programme was the construction of the underground trail.

The fieldwork was carried out from 2001 to 2004 by the team of archaeologists of the Historical and Archaeological Museum in Ostrowiec, under my direction, in close cooperation with miners employed by the firm Geohydrowiert from Kielce and scientists of the Metallurgy and Mining Academy at Cracow. The establishment of the trail marked the end of an important phase of archaeological, mining and conservation work in the Archaeological and Natural History Reserve 'Krzemionki Opatowskie'.

The archaeological and mining project began with the sinking of an exploratory shaft measuring 2-3 m x 9 m (22 m²) halfway between the two tunnels, on the outer extreme of the parabola of the mining field and beyond the limits of the Neolithic flint workshops. In this excavation, a ventilation and evacuation shaft was sunk into solid Jurassic/Oxfordian limestone, beyond the underground workings of the prehistoric mines (Fig. 3). Sunk in 2001, the shaft is 11.2 m deep and reveals the complete geological section of the area (Fig. 4). Three flint-bearing levels were uncovered at the bottom of the shaft, of which the two uppermost ones had been exploited in the Neolithic Period (Fig. 5). The limestone strata incline was calculated at 3° to the south-east (150° off North).

The first gallery, measuring 1.1 x 2 m, was dug in the shaft bottom in solid limestone in the direction of the 'Zenon' shaft and the 'second underground trail' (Fig. 6), taking care to avoid the prehistoric galleries. On the way, four 'windows' were cut which opened only onto the edge of the partly rubble-strewn floor of chamber mines no. 615 and 612. We did not examine any of the underground mines exposed in this section of the trail (the ventilation and evacuation shaft and 'Zenon' shaft), leaving them undisturbed.

Large nodules of flint concretions that had been removed during excavation in the monolithic limestone tourist gallery were weighed: this showed that they are not evenly distributed in layers, but demonstrate local density variations. The differences were as great as 30.5 kg for a length of one metre of the gallery.

Along the course of the trail, some geological disturbances in the limestone strata and some vertical fractures were observed. Dense joint fractures and fault surfaces were also noticed. Joints and faults in the limestone also caused intense fracturing of the flint concretions. There was no point extracting them, given the exceptionally poor

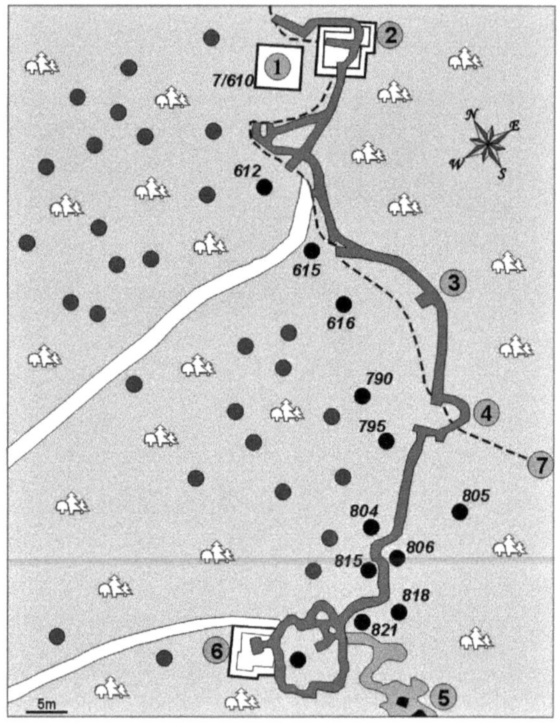

Fig. 3: 'Krzemionki Opatowskie' Reserve, Kielce voivodships. Plan of the underground tourist trail.
1- Neolithic mine unit no. 7/610, 2- Modern 'Zenon' shaft and entry to the underground tourist trail, 3- Modern ventilation and evacuation shaft, 4- Tunnel for tourists, 5- 'Great Chambers', 6- Modern 'Stefan' shaft and exit of the underground trail, 7- Border between solid rock and exploitation field. Black points – prehistoric shafts. 612, 615, 790 etc. – numbers of prehistoric shafts.

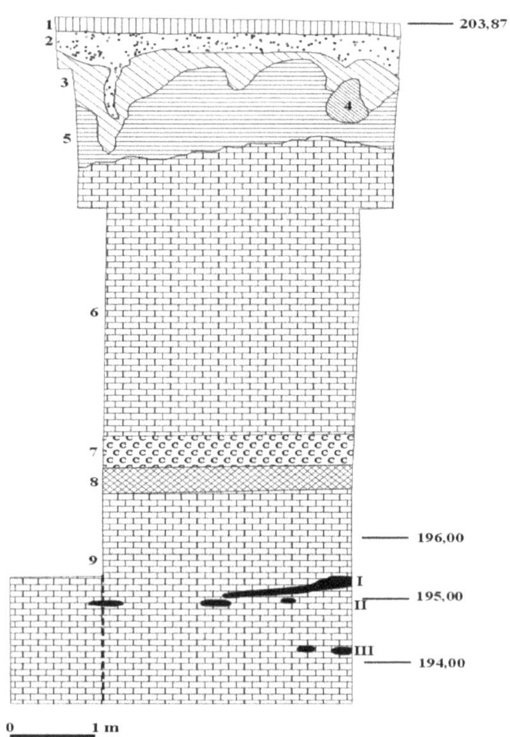

Fig. 4: 'Krzemionki Opatowskie' Reserve, Kielce voivodships. Cross-section of the modern ventilation and evacuation shaft. 1- humus and subsoil, 2- sand, 3,4- clay, 5- lime rubble passing into solid rock, 6, 9- pelitic limestone, 7- coral limestone /limestone with relics of see organisms, 8- oolitic limestone, I,II,III – flint levels.

Fig. 5: 'Krzemionki Opatowskie' Reserve, Kielce voivodships. Flint nodules in situ, at the bottom of modern ventilation and evacuation shaft. Photo: J. Bąbel.

Fig. 6: 'Krzemionki Opatowskie' Reserve, Kielce voivodships. The modern underground tourist trail in the solid limestone rock. Photo: J. Bąbel.

Fig. 7: 'Krzemionki Opatowskie' Reserve, Kielce voivodships. Neolithic mine no. 615. Empty space near the bottom of the shaft. Photo: J. Bąbel.

quality of the material and its lack of any value for knapping. That was indeed the main reason why work had been abandoned in chamber mine no. 615 for example, which showed small galleries and an extremely undulating roof. It turned out to be a 'failed investment' (Fig. 7).

In the course of our work, we reviewed seismic surveys carried out in the 1980s. The hypothetical plan of the Neolithic mines suggested at that time (Borkowski 2000: 254, Fig. 13) does not reflect the actual state of affairs, which seriously challenges the credibility of the seismic method used to survey this kind of construction. In the case of deep Neolithic mines, seismic surveys cannot be relied upon as the basis for archaeological interpretation. The most effective non-invasive subsurface environmental sensing tool is Subsurface Interface Radar surveys.

After linking the gallery with 'Trail 2', we began the work on a similar gallery towards 'Trail 1'.

This section of the trail ran partly through solid limestone and then through the rubble-strewn workings of mines no. 795, 804, 806, 815, 818 and 821 (Fig. 8).

Fig. 8: 'Krzemionki Opatowskie' Reserve, Kielce voivodships. Archaeological underground investigation in the mine no. 795 in 2003. Photo: J. Bąbel.

The first mine mentioned, mine no. 795, is a chamber mine. We chose to investigate only a fragment of its eastern part: two large niches and a smaller one diverge from the main chamber, heading into the depths of the rock mass. We initially estimated the surface area of these galleries at almost 300 m^2. This probably amounts to less than half of the underground chamber of mine no. 795, making it the largest known Neolithic flint mine in Poland. Ventilation is another extremely important issue. It is generally accepted that the maximum direct distance from the bottom of the shaft to the surface should not exceed 12 m. Since a high concentration of carbon dioxide builds up in the area where a miner is working at the rock face, this can cause ventilation problems. In the case of low passages - up to a height of around 1 m - where torches were used, this concern was of fundamental importance.

Fig. 9: 'Krzemionki Opatowskie' Reserve. Mine no. 815. Cross-section of the fireplace in Neolithic waste-heaps. Photo: J. Bąbel.

In chamber mine no. 795 the direct distance from the bottom of the shaft to the furthest face was around 20 m. This face was located at the outside boundary of the mining field, in a place where the flint-bearing layer was dipping into the centre of the flint syncline. In that case, suitable ventilation was naturally present. Air warmed by the presence and activity of human bodies travelled along the ceiling towards the shaft, while cold fresh air travelled just above the floor of the gallery. This system worked best in winter, when there were large temperature differences between the surface and the underground. Please recall that the temperature underground in Krzemionki is a constant 6-9 °C throughout the year.
A great deal of charcoal (the remains of wooden torches which provided light during mining) was found during excavation. Its presence also indicates ventilation. In order to create an air flow, several torches had to be burned at the mine face and in the tunnels. An additional means of intensifying the air flow was to keep a fire burning at the base of the shaft. Vestiges of such a fire, in the form of a 10-15 cm level of charcoal, were discovered near the spoil heap of shaft no. 815 (Fig. 9).

Fig. 10: 'Krzemionki Opatowskie' Reserve, Kielce voivodships. Mine no. 815. Neolithic charcoal drawing on the wall. Photo: J. Bąbel.

Torches were made of tree branches of 3-4 cm in diameter. In order to keep the flame alight, the miner would rub it against the sharp edge of limestone slab, a piece of waste, or against the sidewall of the mine. This leaves behind a characteristic pattern, a lattice of numerous criss-crossed lines. Numerous examples of this phenomenon were observed in the chamber of mine no. 795, which was investigated. In some cases

Fig. 11: 'Krzemionki Opatowskie' Reserve, Kielce voivodships. Mine no. 804. The wooden torch in its original position. Photo: J. Bąbel.

Fig. 12: 'Krzemionki Opatowskie' Reserve, Kielce voivodships. Mine no. 795. Neolithic communication gallery. Photo: J. Bąbel.

Fig. 13: 'Krzemionki Opatowskie' Reserve. Mine no. 795. Neolithic technological gallery. Photo: J .Bąbel.

the irregular lines of the marks are arranged in more regular patterns, which may suggest some symbolic significance (Fig. 10). Future research will test this initial hypothesis.

A unique find was a well-preserved fragment of a Neolithic torch which had escaped decay (Fig. 11). This was discovered at the boundary between mines no. 815 and 804. C14 dating of part of the torch would place it in the period between 3330 and 2907 BC (Bąbel *et al.* 2005: 542).

In the underground galleries of mines no. 795 and 804, connecting and so-called 'communication and technological' tunnels were discovered (Fig.12, 13). The gallery in one place revealed a so-called 'node', while in another - mine no. 804 - it cut straight across a Neolithic gallery. In no case does the most recently dug tourist trail run along the prehistoric gallery, but through solid rock or through Neolithic spoil heaps.

The gallery uncovered in mine no. 795 was made up partly of the connecting corridor running from the bottom of the shaft to the 'node' and partly of the so-called 'technological corridor'. The purpose of the former was mainly to haul out the excavated material. The latter led from the node to the face uncovered in window no. 4. During the course of the mine's exploitation, the course of that gallery was modified. The connecting corridor in mine no. 795 possesses smoothly polished sidewalls and floor. This contrasts with the 'technological corridor' built in the spoil heap constructed of sharp-edged limestone nodules.

Up to now, it had been thought that the flint mined from the rock was hauled in woven wicker baskets. With regard to mine no. 795, this cannot have been the case, as the haulage vessels left no clear marks or scratches on the sidewalls or floor of the connecting gallery. A significant difficulty for Neolithic miners would have been the tendency of such baskets to get stuck in the narrow passages, or caught on the protruding edges of limestone lumps, causing the gallery bottom to be quickly covered. The baskets themselves would have quickly been destroyed. We suggest then that the raw material used for the production of containers was thick animal hide. Such leather sacks, dragged on the fur side, would have been easy to haul. The strong, flexible material would have been only slightly deformed on bends and other obstacles, and would have slid smoothly over uneven surfaces. This hypothesis should be investigated in the future by an examination of micro traces by specialists.

The gallery in column mine no. 804 leads under the rock walls dividing up the chamber and further along the sidewall to the furthest part of the mine. The sidewalls of this gallery

Fig. 14: 'Krzemionki Opatowskie' Reserve, Kielce voivodships. Mine no. 795. Cross-section of the prehistoric underground heap of limestone rubble used as filling material. Photo: J. Bąbel.

Fig. 15: 'Krzemionki Opatowskie' Reserve. Mine no. 795. Cross-section of prehistoric underground waste-heaps (dry filling). Photo: J. Bąbel.

Fig. 16: 'Krzemionki Opatowskie' Reserve, Kielce voivodships. Mine no 795. Negative of a wooden tool in Neolithic waste-heaps (dry filling). Photo: J .Bąbel.

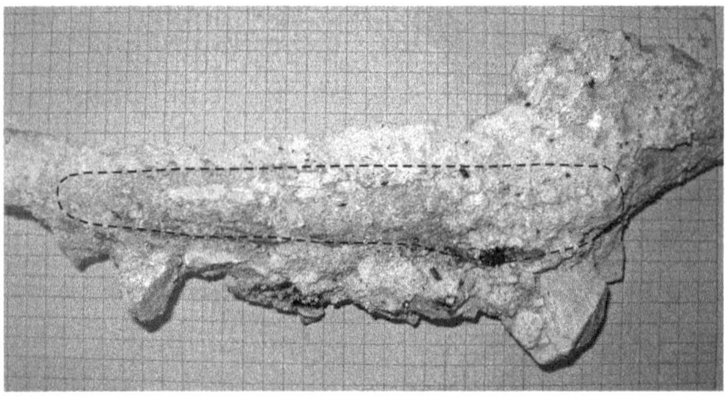

Fig. 17: 'Krzemionki Opatowskie' Reserve. Mine no. 795. Positive of a wooden tool (in gypsum). Photo: J. Bąbel.

Fig. 19: 'Krzemionki Opatowskie' Reserve Kielce voivodships. Mine no 795. Fragments of the mines stone pick. Photo: J. Bąbel.

Fig. 18: 'Krzemionki Opatowskie' Reserve. Mine no. 806. Negative of a wooden construction in Neolithic waste-heaps (dry filling). Photo: J. Bąbel.

are not polished and its floor was strewn with large fragments of limestone waste. Most probably this corridor served as a connecting artery, linking shaft 804 with the other galleries adjacent to this mine from the east.

The tourist tunnel that we dug out cut across several underground Neolithic spoil heaps (Fig. 14, 15), which, after extensive documentation, were suitably protected from destruction and collapse. The perfectly visible artificial bedding produced in the limestone waste during its collection makes possible the full reconstruction of the sequence of work of Neolithic miners in that part of the mining area. This issue will be the subject of a separate paper.

During the exploration of the underground spoil heaps, we came across small hollow spaces which were impressions made by wooden tools and objects (Fig.16). Rotten remains strewn about attest to this. Rock dust and limestone waste sticking to them preserved the shapes of these objects. The cylindrical hollow spaces were around 50-70 cm in length with a diameter of 3-4 cm. They were probably the remains of levers used to prise off large slabs from the roof of the mine. A plaster cast was made of one of these finds. They must also have been used as wedges to split the rock (Fig. 17). The impression of one such tool was 19 cm long around 6 cm wide and up to 2 cm thick.

The impressions were sometimes found together in groups. In two places some distance apart from each other, in mine no. 804, we came across several impressions of wooden poles arranged vertically and horizontally. They were 1 m long and 2-9 cm in diameter. In one place they were clearly located at the border of two layers of limestone waste, between a spoil heap and a workface later buried in waste. This suggests that various wooden constructions were built to protect the gallery from the spoil heaps collapsing, with the result that the gallery became narrower and moving through it became more difficult. In places where the particular type of rock mass split into small cubes, it was impossible to produce large limestone slabs to line the corridor, and primitive timbering was put up instead (Fig. 18).

Plant remains, which may have derived from woven mats, have been found among the waste in underground Neolithic spoil heaps. Plaited mats may have also been used during mining to insulate the body of a crouching or prone miner from the cold and damp rock floor. This would have been a natural measure to prevent such illnesses as rheumatism, gout or acute lumbago etc.

Other mines, encountered during the connection of the tourist trails, much smaller and shallower than mine no. 795, can be categorised as chamber and pillar mines.

The examination of the underground spoil heaps also revealed well-preserved fragments of stone and flint tools discarded by miners (Fig. 19). The majority of them are cigar-shaped picks made of crystalline rock, and hammerstones. The latter were mainly used as hammers to test the quality of the raw material extracted from the rock. We also found wedges made of fragments of polished axes, characteristic of the Globular Amphora culture. Another group of tools made of fragments of red deer, roe deer and elk antlers were used as wedges, levers, picks and gouges (Fig. 20, 21). A spade made from an animal's shoulder blade, found in one of the work faces of mine no. 795, was probably used to rake up and scatter fine waste.

On the mine sidewalls, and particularly on work faces in niches, thousands of marks left by the tools of Neolithic miners can be seen (Fig. 22). These marks provide us with a great deal of information about the miners themselves. They were right-handed. They would change

Fig. 20: 'Krzemionki Opatowskie' Reserve Kielce voivodships. Mine no. 795. Neolithic antler tool in its original position (in situ). Photo: J. Bąbel.

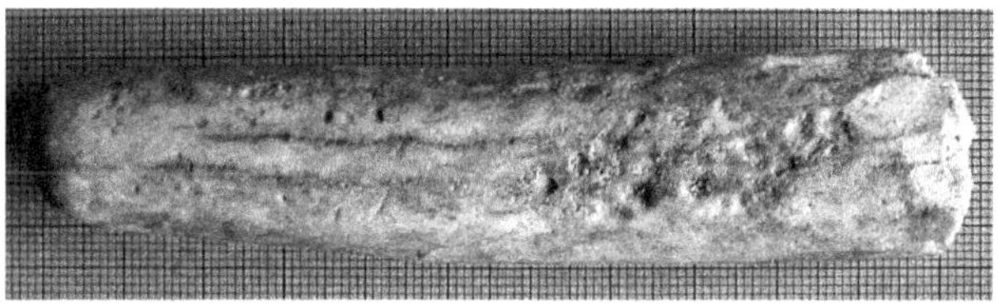

Fig. 21: 'Krzemionki Opatowskie' Reserve Kielce voivodships. The antler wedge. Photo: J. Bąbel.

Fig. 22: 'Krzemionki Opatowskie' Reserve, Kielce voivodships. Mine no. 805. Traces of works on the wall of Neolithic gallery. Photo: J. Bąbel.

Fig. 23: 'Krzemionki Opatowskie' Reserve, Kielce voivodships. Mine no 815. Traces of work of an antler gouge. Photo: J. Bąbel.

tools, depending on the work being carried out at the time and on the type of rock they had to deal with in a given location, in a remarkably rational way. The best examples are the marks of horn wedges used to prise off large pieces of limestone on the roof of mine no. 815 (Fig. 23). Gouges used there left indentations of their blades in the clay which filled a karst crevice. The technique of extracting flint nodules from rock is illustrated by a series of marks in one of the faces in mine no. 795 (Fig. 24). There, marks of impacts can be seen which split off the concretions from the rock sidewall, as well as the mark of a lever which split flint away from the floor. In mine no. 795, marks of where large limestone slabs were broken off from the roof of the workings can be seen (Fig. 25). The size of these slabs and their weight - which reached as much as several hundred kilograms in some cases - force us to consider how they were broken off in such cramped and low passages.
Large accumulations of flint flakes together in a niche of pillar mine no. 818 argue for the existence of an underground workshop for preliminary flint knapping, to test the stone's quality. Other places like these were discovered in the region of Trail 1 in 1983.

Fig. 24: 'Krzemionki Opatowskie' Reserve. Mine no 795. Negative of a flint nodule. Photo: J.Bąbel.

Fig. 25: 'Krzemionki Opatowskie' Reserve. The mine no. 795. Traces of works of a wooden wedge. Photo: J.Bąbel.

Before the flint reached the workshop on the surface next to the shaft, its use value was assessed. Micro cracks, formed as a result of tectonic movements of the rock mass, make up part of any normal concretion. Neolithic miners, working underground by the light of a torch, would obviously not have been able to see such cracks; nonetheless, most poor quality flint remained below the surface. In my opinion, the sense of hearing rather than of sight played a crucial role in testing flint. As he worked, the miner not only saw the pieces of flint breaking and falling under blows of his hammerstone, but he also heard whether the sound was clean or dull. In the latter case, the flint was of no further use and ended up along with limestone waste on the underground spoil heap.

Fig. 26: 'Krzemionki Opatowskie' Reserve, Kielce voivodships. Mine no 795. Traces of water filling on the limestone rubble in the chamber-mine. Photo: J. Bąbel.

The horizontal watermark of a temporary underground lake can be seen on the walls of Neolithic galleries built from limestone slabs, and on the sidewalls of the deepest part of mine no. 795. It must have arisen as a result of heavy rains before the shaft was buried under waste. Rainwater entered the underground from the surface through the open shaft (during the Neolithic Period?) carrying in

Fig. 27: 'Krzemionki Opatowskie' Reserve, Kielce voivodships. Mine no. 795. The underground tourist trail is ready. Photo: J. Bąbel.

with it various soil components, mud, flecks of humus and of rotten vegetation. Before the water soaked into the rock, the scum on its surface left its mark in the form of a thick line (Fig. 26). Identical marks can be seen on the Neolithic sidewalls of the galleries in the 'Great Chambers'. These observations may interest climatologists researching climate changes in past epochs.

During the excavation, numerous samples of charcoal were extracted from the part of the trail running through chamber mines and chamber-and-pillar mines. Thanks to a collaborative agreement with the Holy Cross Academy of Kielce and the University of Erlangen in Germany, we obtained 10 new C 14 datings (Bąbel *et al.* 2005). After calibration, these gave readings of around 3500 - 2900 BC. Previously, it had been thought that chamber mines were the work of miners of the Globular Amphora culture. These dates however are inconsistent with the generally accepted chronology for that culture. This means that the question of during which culture these mines were exploited remains controversial.

A long stretch of the tourist trail runs beyond the mining area along a gallery dug out of solid limestone on whose sidewalls the section of flint-bearing layers can be seen. Observations of these sections and planigraphy aided geologists from the Polish Geological Institute to propose a new hypothesis for the genesis of Jurassic (Oxfordian) banded flint (Pieńkowski and Gutowski 2004).

The tourist trail has been made completely safe for visitors (Fig. 27). Traditional support techniques have been used, along with pressure injection methods. In addition, Neolithic man-made spoil heaps have been fenced off from the public by steel mesh hung between columns to prevent undesirable interference.
The underground exhibition gallery, about 480 m long, passing through Neolithic mining units, was opened for tourists on 1 July 2004.

The Krzemionki underground tourist trail has great educational value. In 2004-2006, over 160 000 people visited it, making it a record in the history of the museum and archaeological reserve.
In May 2005, the Minister of Culture awarded the Historical and Archaeological Museum in Ostrowiec Świętokrzyski the third prize in the category of archaeological exhibitions in the competition of Museum Events - Sybilla 2004, for its underground tourist trail created in the Neolithic flint mines of Krzemionki.
And in October 2006, the Minister of Economy awarded our Museum the Certificate from the Polish Tourist Organisation – 'Best tourist product of 2006 Travel with passion year' for the underground tourist route in prehistoric flint mines and the reconstruction of the Neolithic village' in the 'Krzemionki' reserve.
Another extremely important benefit is that the tunnel has opened space for future specialised research. This, with the help of improved techniques and technology, will significantly improve the current understanding of the prehistory of the Krzemionki mines.

Dr Jerzy T. Bąbel
ul. M. Dabrowskiej 19m9
01-903 Warszawa POLAND.
e-mail: tbabel@poczta.onet.pl
website: www.krzemionki.pl

REFERENCES

Bąbel, J.T. 1975. Zniszczenia, badania i ochrona rezerwatu w Krzemionkach. Wiadomości Archeologiczne 40, 149-177.

Bąbel, J. 1980. Krzemionki. In G. Weisgerber, R. Slotta, and J. Weiner (eds), 5000 Jahre Feuersteinbergbau. Die Suche nach dem Stahl der Steinzei, Bochum, 86-595..

Bąbel, J. 1983. Krzemionki dziś i jutro. Stan, badania i perspektywy zagospodarowania rezerwatu archeologicznego. Wiadomości Archeologiczne 48, 223-236.

Bąbel, J. 1986. The Problems of investigations of the flint mine at Krzemionki, near Ostrowiec Świętokrzyski'. In K.T. Biro (ed.) International Conference on Prehistoric Flint Mining and Lithic Raw Material Identification in the Carpathian Basin, Budapest - Sümeg, 20-22 May 1986, Budapest, 27-42.

Bąbel, J. 1990. The flint mine at Krzemionki and the problem of flint workshop from the Early Bronze Age in the Central Little Poland. In M.-R. Seronie - Viven and M. Lenoir (eds), Cahiers du Quaternaire no 17 - Le silex de sa genèse l'outil, Actes du Vème colloque international sur le silex. Bordeaux, 201-209.

Bąbel, J., Braziewicz, J., Jaskóła, M., Kretschmer, W., Pajek, M., Semaniak, J., Scharf, A. and Uhl, T. 2005. The radiocarbon dating of the Neolithic flint mines at Krzemionki in Central Poland. Nuclear Instruments and Methods in Physics Research 240, 539-543.

Balcer, B. 1995. The relationship between a settlement and flint mines. A preliminary study of the Eneolithic workshop assemblages from Ćmielów (Southern Poland)'. In Z. Kobyliński, Lech J. (eds), Archaeologia Polona 33, 209-221.

Balcer, B. 2002. Ćmielów – Krzemionki- Świeciechów. Związki osady neolitycznej z kopalniami krzemienia. Warszawa.

Balcer, B., and Kowalski, K. 1978. Z badan nad krzemieniem pasiastym w pradziejach (Sum.: Studies on striped flint in prehistory). Wiadomości Archeologiczne, 43(2), 127-145.

Barber, M., Field, D., and Topping, P., 1999. The Neolithic Flint Mines of England. London.

Borkowski, W. 1990. Results of subsurface radar geophysical studies of the Krzemionki banded flint mines Poland. Archeometry 90, 687-696.

Borkowski, W. 1995a. Krzemionki Mining Complex: deposit Management System. Warsaw

Borkowski, W. 1995b. Prehistoric flint mines complex in Krzemionki (Kielce Province). In Z. Kobyliński and J. Lech (eds), Archaeologia Polona 33, 506-24.

Borkowski, W. and Budziszewski, J. 1995. The use of stroped flint in Prehistory. In Z. Kobyliński, and J. Lech (eds), Archaeologia Polona 33, 71-87.

Borkowski, W. 2000. Zastosowanie metody radarowej i sejsmicznej w lokalizacji podziemnych wyrobisk górniczych na podstawie wyników prac prowadzonych w Krzemionkach Opatowskich. In W. Borkowski (ed.), Metody badań archeologicznych stanowisk produkcyjnych – górnictwo krzemienia. Warszawa.

Budziszewski, J. and Michniak, R. 1983(89). Z badań nad występowaniem, petrograficzną naturą oraz prahistoryczną eksploatacją krzemieni pasiastych w południowym skrzydle niecki Magoń – Folwarczysko. Wiadomości Archeologiczne 49, 151-189.

Engelen, F.H.G. 1980. Rijckholt-St.Geertruid, Prov. Limburg. In G. Weisgerber, R. Slotta and J. Weiner (eds), 5000 Jahre feuersteinbergbau. Die Suche nach dem Stahl der Steinzeit, Bochum 559-567.

Felder, P.J. (Sjeuf), Rademakers, P., Cor, M. and de Grooth, M. E. Th. (eds.) 1998. Excavations of Prehistoric Flint Mines at Rijckholt - St. Geertruid (Limburg, The Netherlands) by the 'Prehistoric Flint Mines Working Group' of Dutch Geological Society, Limburg Section. Archaeologische Berichte, 12, Bonn.

Hadamik, Cz. 2006. Krzemionki- Flint Mines from the Neolithic Period. In M. Róziewicz and J. Wendland (eds), Historic Monuments in Poland, 30 Treasures of National Heritage, Warszawa, 140-146.

Herbich, T. 1993. The method of estimation of the extent of the mining field of flint mines through observation of the arrangement of surface layers. Archeologia Polski 38, 23-35.

Lech, J. 2004. O badaniach prehistorycznego górnictwa krzemienia i kopalni w Krzemionkach Opatowskich. Przegląd Archeologiczny 52, 15-88.

Pieńkowski, G. and Gutowski, J. 2004. Geneza krzemieni górnego oksfordu w Krzemionkach Opatowskich, Genesis of the Upper Oxfordian flint in Krzemionki Opatowskie, Poland. Tomy Jurajskie 2, 29-36.

OPEN-CAST FLINT MINING, LONG BLADE PRODUCTION AND LONG DISTANCE EXCHANGE: AN EXAMPLE FROM BULGARIA

Laurence Manolakakis

Abstract: In the early 5th millenium, a vaste area termed 'Graphite Pottery Zone' in the Balkan Peninsula underwent radical transformation: emergence of hereditary social hierarchy, specialization and regionalization of productions, highly organized long distance exchange, and technological innovations. A particular aspect of the latter was the specialized production of very long flint blades. In north-east Bulgaria, Ravno near Razgrad, is a good-quality flint deposit, and the tell of Kamenovo is a characteristic village of the Graphite Pottery Zone, with two major differences: the village is located on a flint deposit and the inhabitants produced long blades by lever pressure as prestige items. The raw material procurement of Kamenovo combines exploitation of flint on the site for domestic production and further exploitation a short distance away (Ravno) of high quality raw material for specialized production. Settlement location away from the Ravno source may reflect the fact that Ravno was probably exclusively exploited by specialists, maybe as a protected area, and that long blade debitage was not an intensive production. One may suggest that intensive exploitation developed when an object was not or no longer a prestige item.

Keywords: Open-cast flint mining, domestic production, specialization, exploitation, procurement, Eneolithic, Bulgaria.

1. CULTURAL-HISTORICAL CONTEXT

From the early fifth millennium BC, certain populations in Europe underwent radical technological, economic and social changes. The term Chalcolithic is usually applied to these upheavals, which included the establishment of durable social hierarchy, the emergence of virtually professional craftsmen, and the development of long-distance networks for circulation and exchange of goods. The cradle of this transformation was the Balkan Peninsula, with a major cultural area called the 'Graphite Pottery Zone' covering the whole of Bulgaria, the Romanian Danube plain, the Republic of Macedonia and the north coast of the Aegean. It lasted from about 5000 to 4300 BC and is divided into two main periods. According to the terminology used in Bulgaria, based on the tell of Karanovo, these are the early and late Eneolithic, corresponding to the late Neolithic and early Chalcolithic of the pan-European terminology (Lichardus *et al.* 1985). While the end of this era is poorly dated and documented, there is abundant evidence throughout the first half of the fifth millennium, allowing definition of a multitude of cultural groups, cultures, regional groups and cultural complexes (Todorova 1978). Yet the various cultural groups which made up the Graphite Pottery Zone share a number of features, particularly flint production.

A clearly differentiated, hereditary social hierarchy, which first appeared in the early Eneolithic, was completely established in the late Eneolithic, as reflected in the Varna cemetery (Ivanov 1978). In the early Eneolithic, copper metallurgy not only involved the manufacture of small objects but also of heavier tools such as axes and hammer-axes. In the late Eneolithic gold-work emerged, for

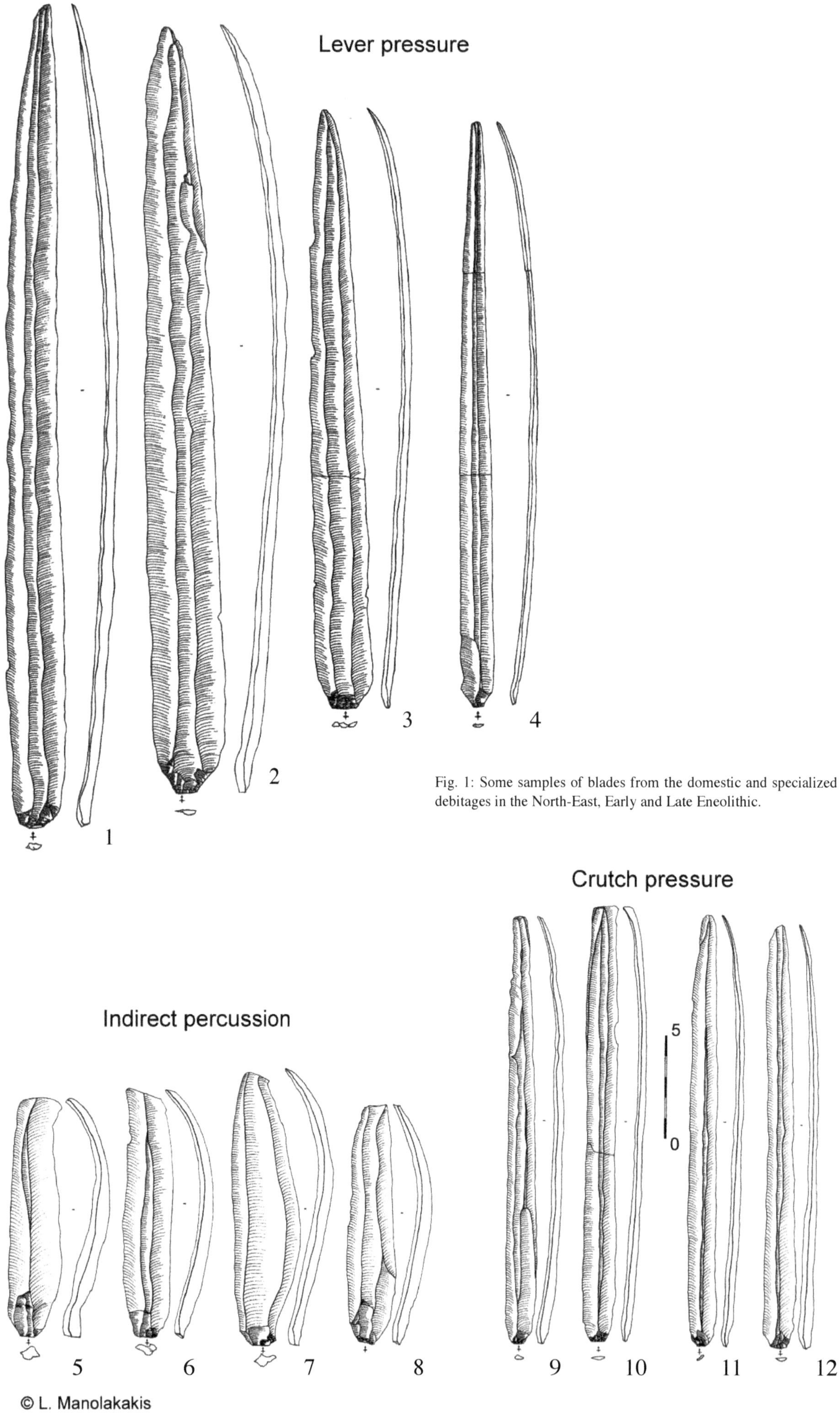

Fig. 1: Some samples of blades from the domestic and specialized debitages in the North-East, Early and Late Eneolithic.

both ornaments and symbols of power, particularly sceptres. The whole Eneolithic is also characterized by high-temperature firing of pottery decorated with graphite, or with gold in the cemeteries, by lever pressure debitage, and long-distance circulation networks for certain goods such as copper and very long flint blades.

Production was organized regionally, in accordance not only with the presence of raw materials – copper, graphite, spondylus shell, gold, flint – but also with technical know-how, as shown by the various production centres for very long flint blades (Manolakakis 2005), gold ornaments (Angelov 1959, Hartmann 1978), copper tools (Tchernih 1978a & b), bone figurines (Angelov 1958, 1961) and probably graphite decorated pottery as well (Demoule 2004).

2. LITHIC PRODUCTIONS

Lithic production using flint can be summarized as follows (Manolakakis 2005) (Fig. 1 n°5-8): domestic blade debitage was carried out within each village, using local flint. This occured in all regions and represented a large majority of cases (90% of lithics on a site). Products did not circulate between villages, or between regions. Debitage differed from one region to another. Within each of the five regions, the same debitage was used, indicating that all settlements shared the same technical knowledge and conception of debitage. In the current state of research, there is no evidence for specialized production within the domestic sphere. Regional domestic production can be considered traditional, as it was already present in the early Eneolithic and remained unchanged in the late Eneolithic. A roughly east-west separation is visible, between the north-east and Thrace on the one hand and west Bulgaria on the other. Briefly put, indirect percussion is used in the eastern area and direct percussion in the west. Due to its considerable uniformity, the eastern half of Bulgaria (north-east and Thrace) formed a large unit matching the cultural complex defined on the basis of ceramics, the Boian-Marica of the early Eneolithic and the Kodzadermen-Gumelnitsa-Karanovo VI of the late Eneolithic, both belonging to the Graphite Pottery Zone.

At the same time, there is specialized debitage: throughout the periods, there is evidence from all the villages in the eastern half and some in the western half of a few blades which do not originate from domestic debitage, and for which there is no knapping waste present. These blades always represent less than 12% of the lithic industry from villages, but they are much more common as grave-goods in cemeteries, where they represent as much as 40 to 68%. This is mainly lever pressure debitage of very long blades, from 22 to 45 cm long. To a lesser extent, crutch pressure debitage occurs, of 'shorter' blades, roughly between 16 and 20 cm long (Fig. 1 n°1-4 and 9-12). These objects circulated between villages as semi-finished products, up to a distance of 700 km .

During the late Eneolithic, though only in the eastern zone, axes made from very long blade cores also circulated, as finished products and in smaller quantity (Fig. 2).

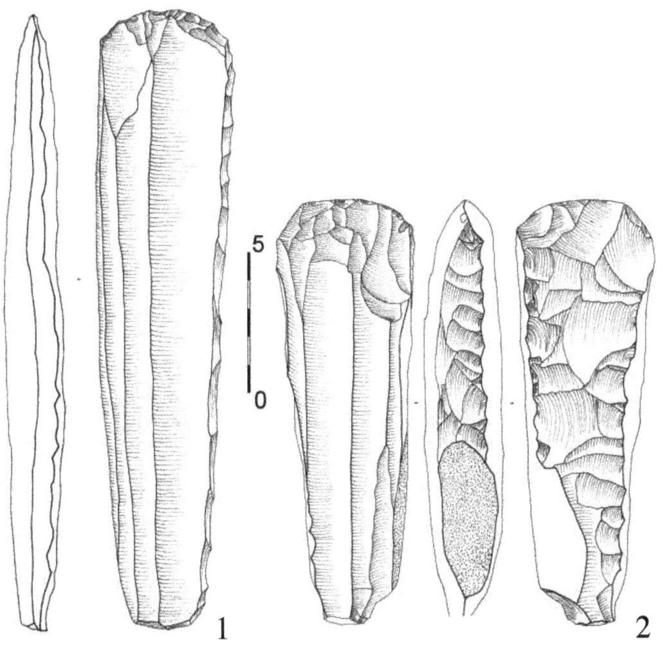

© L. Manolakakis

Fig. 2: Flint axes made on pressure lever cores, only in the Late Eneolithic.

3. DEPOSITS AND RAW MATERIAL EXTRACTION

The raw material used for this pressure debitage is exclusively Aptian flint from the so-called 'pre-Balkanic platform' in north-east Bulgaria (Natchev *et al.* 1981, Natchev, Kantchev 1984) (Fig. 3). There, this flint was also used for domestic

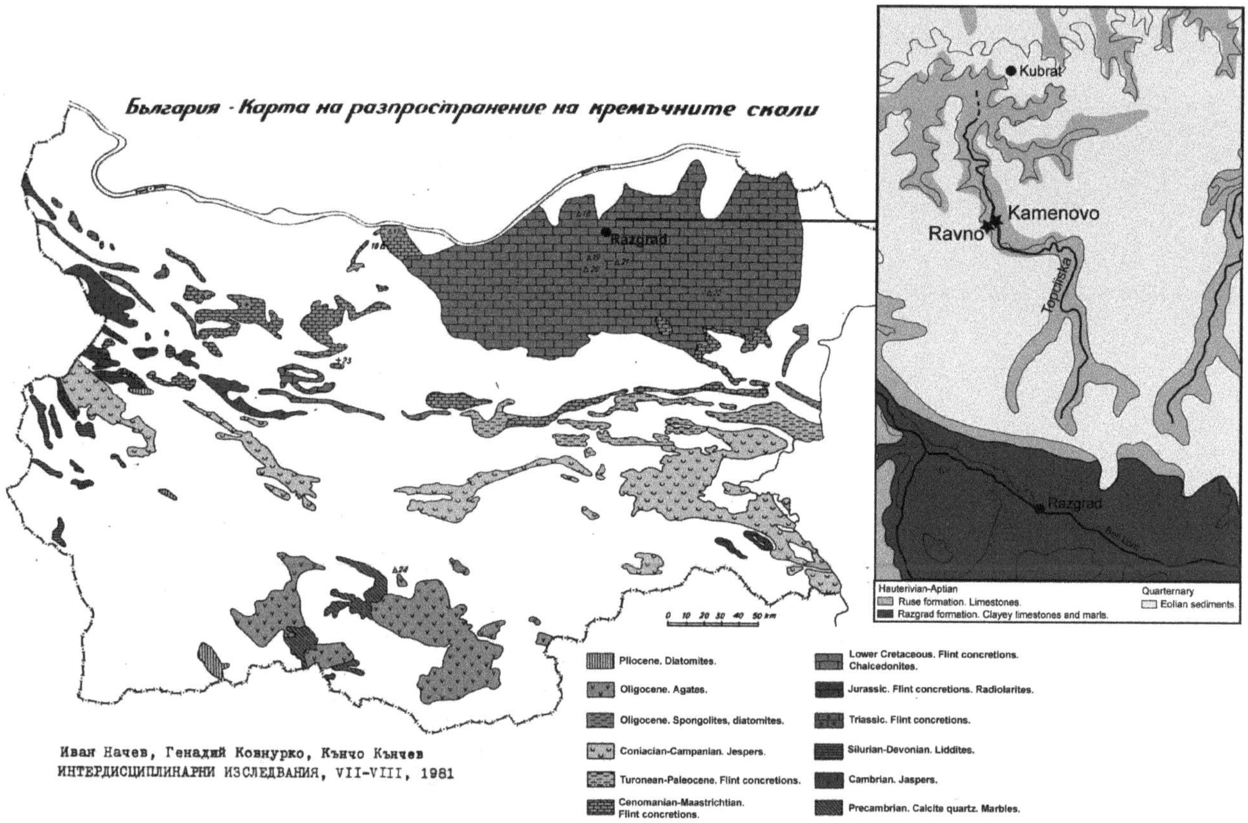

Fig. 3: Map of the siliceous rocks of Bulgaria (from Natchev *et al.* 1981) and geological location of the study area.

debitage. While all the north-eastern villages carried out indirect percussion debitage on local flint, only a few used pressure debitage as well. However, there was until recently no evidence either for deposits containing flint of an appropriate size for debitage of long blades, nor for production sites. Debitage of very long blades, of course, implies nodules not only of good quality but also of large dimensions.

Lower Cretaceous flint is abundant throughout the Moesian plateau, where Hauterivian and Barremian flint containing deposits are the most frequent. But there are also small concentrations of weathered Aptian flint, especially north of Razgrad (Fig. 3) (J. Delépine, in unpublished excavations report 1996).

This is why geological and archaeological survey of the region between Kubrat and Razgrad was undertaken from 1994 to 1997 in collaboration with Ivan Ivanov (Museum of Varna) and Jacques Delépine (geologist). The two main objectives of the project were to locate flint deposits with nodules large enough for lever pressure debitage and to find archaeological sites with waste from this kind of debitage.

Thus in the small Topciiska valley, a dozen geological trenches up to 5 m deep were opened, and a surface-area of 543 ha surveyed.

Three deposits were located at Ravno, containing abundant raw material. The interest of this area was further enhanced by the presence of a late Eneolithic settlement at Kamenovo (Fig. 4).

'Ravno 3' is the only deposit which provides nodules suitable for debitage of long blades; the site is part of a regional system of dolines and lies on a slope which flattens out towards the bottom. Four trenches were lined up from the summit down to the bottom, enabling complete coverage of geology and sediments. They were 3-15 m long, 1 m wide with a depth ranging from 3-5 m. The stratigraphic section of the trench at the bottom of the slope (Fig. 5) shows:

- At the top, superficial silts.
- At the base, the Lower Cretaceous limestone bedrock, dating to the Hauterivian and the Barremian, which can be up to 500 m thick.
- Between the two are the bright red or yellow clays of the residual, weathered Aptian layer, containing large numbers of flint nodules.

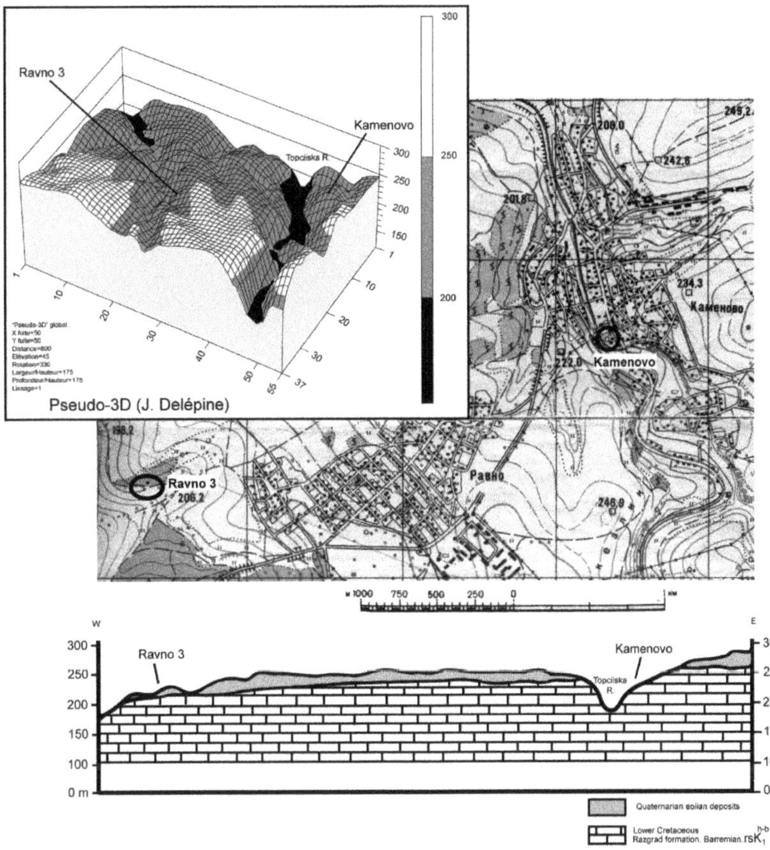

Fig. 4: Topographic location of Ravno deposit and the tell of Kamenovo (Delépine, unpublished excavations report 1996).

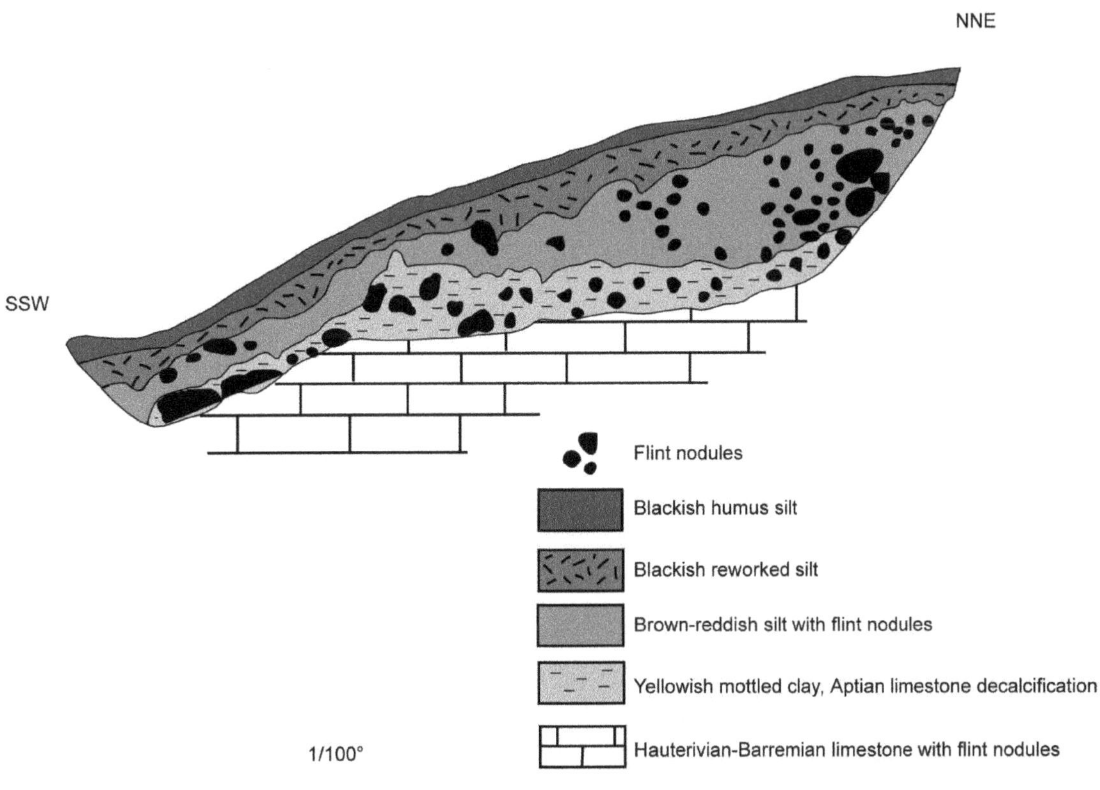

Fig. 5: Stratigraphical profile of the lowest geological trench of Ravno 3 deposit (Delépine, unpublished excavations report 1996).

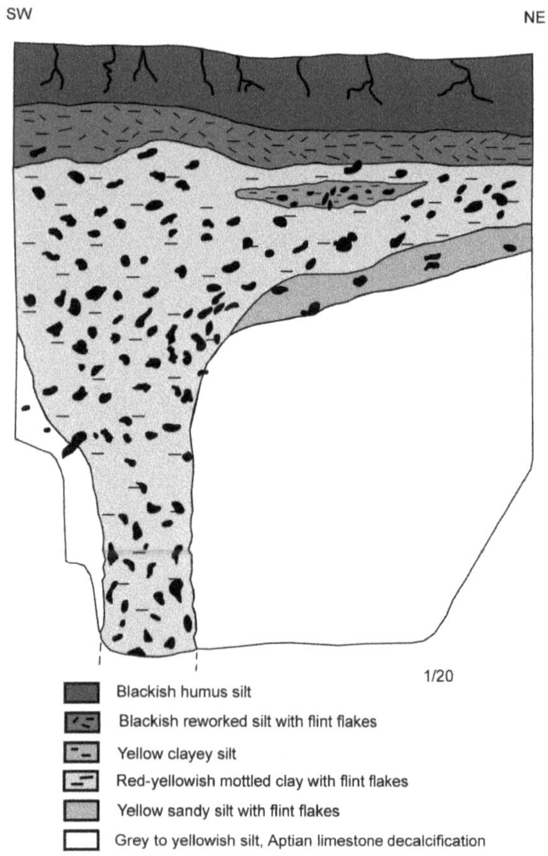

Fig. 6: Stratigraphical profile of the shaft of Ravno 3 flint deposit (Delépine, unpublished excavations report 1996).

The nodules, of good quality, are 40 to 80 cm long and occasionally reach 1 m. They are globular in shape, sometimes with small surface lumps. The fresh flint is brown to beige in colour and the cortex is non-effervescent white.

An archaeological feature was discovered in the trench at the bottom of the slope. The shaft, which appeared just below the topsoil, is funnel-shaped with a wide opening at the top (Fig. 6). At a depth of about 2.5 m, the feature rapidly becomes a round shaft with an average diameter of 0.60 m. The depth is over 4.5 m, since the base was not reached. So narrow a feature suggests that this was an access to a gallery in the flint bed. The bed, located at a depth of 2.50-5 m, depending on the position on the slope, contained large numbers of quite large flint nodules. The fill of the shaft consisted of speckled clayey silt containing a great quantity of worked flint, although there is no waste from lever pressure debitage.
The untouched nodules are all smaller than 12 cm. Large numbers of nodules had been tested but rejected because of poor quality: frost damage, cleavage or other defects. The test was made with direct percussion with a hard hammer (probably in stone). In most cases, the shape of the nodule meant that the first flake also created the future striking platform. In other cases, the nodule was tested by removing flakes on two opposite sides, using a natural striking platform. Raw material of poor quality was immediately identified by most tests. However, some examples indicate that an attempted preparation had been abandoned due to cleavage within the nodule. Most cores were pyramidal with flat striking platforms and a wide debitage surface, 15-18 cm long. Debitage was of long, wide blades, curved and varying in regularity. The cores were shaped by means of crests, often leaving the back more or less cortical. There were also cone-shaped cores, with very oblique and flat striking platforms, narrow debitage surfaces and pointed bases. They are 10-15 cm long and indicate debitage of medium-sized, slightly curved regular blades. There were considerable numbers of large first flakes, shaping flakes, crested blades, core rejuvenation tablets, debitage splinters, debitage flakes, laminar flakes for renewing debitage surfaces, and crested flakes. Most blades had flat butts, abrased overhangs, diffuse bulbs, two or three regular or irregular arrises, and generally a curved profile. These blades are quite short (9 cm). The longer ones (14 cm) are in fact laminar flakes from debitage surface renewal (correcting hinged fracture). The blades are often fragmentary. The finds from the shaft thus present all stages of debitage, probably carried out by indirect percussion, and logically consist almost entirely of waste from these stages. The shaft was indeed used to extract flint. The preliminary stages of shaping were carried out by direct percussion with a hard hammer, then full debitage by indirect percussion. This is no evidence to date the site to the late Eneolithic, as punch percussion debitage also took place during the following period (Bronze Age according to Bulgarian terminology), and no pottery was found. One can also note that the cores and waste from the shaft are much less regular than on Eneolithic sites in the north-east. Additionally, in the late Eneolithic, indirect percussion debitage was always carried out in the settlements themselves. Since there are no traces of settlement at Ravno, this would suggest that the material is posterior. In this situation, it was important to investigate the small late Eneolithic tell of Kamenovo, about 2 km away, especially as it lies on a large flint deposit.

4. THE TELL OF KAMENOVO

The tell is located next to the Topciiska, at an altitude of 226 m in the same geological context as Ravno (Fig. 4). Large quantities of good-quality flint are available on the site, but the nodules are smaller than at Ravno 3. In 1999, in collaboration with Ivan Ivanov and Aline Averbouh, two perpendicular trenches were opened, 20 m long and 2 m wide (Fig. 7, trenches A, B). These trenches could not be placed on the top of the tell as this is occupied by the present-day village. They were thus located in the only accessible area (a public garden), on a lower-lying part of the tell. The project was interrupted after the death of Ivan Ivanov, but provided interesting results. Most interpretation of the data remains hypothetical because of the limited scale of the excavations. The basic stratigraphy can be described as follows (Fig. 7):

- Topsoil: 0.20-0.40 m thick, clayey brown, containing some Eneolithic ceramics and lithics, as well as a few objects from the Thracian period and a great many from the 19th and 20th centuries.
- Level 0: 0.80-1.10 m thick, light grey, clayey silt with some grey-brown lenses of similar texture. This level was disturbed and the archaeological finds here are out of context. Micro-sediment analysis should determine whether this was a backfill horizon. The presence of this disturbed level is perhaps due to the location of the trenches quite low down on the side of the tell. Although there were clearly no features in this disturbed level, finds were very abundant and are all datable to the Eneolithic.
- Level 1: appeared at 1.60-1.80 m depth, fine clayey grey. This first undisturbed level was only found in the north-east/south-west trench (about 18 square m, trench A) and produced two archaeological features. Feature 1 was visible for only 0.60 m within the excavation trench. It was a roughly rectangular mass of burned daub, apparently part of the wall of a collapsed building. Similarly, feature 2 was visible in the section and for 0.80 m within the trench. This too was a more or less rectangular mass of burned daub and probably also a fragment of collapsed wall. As both features were mainly seen in the profile, it was not possible to determine whether they corresponded to a single structure or to two adjacent buildings.

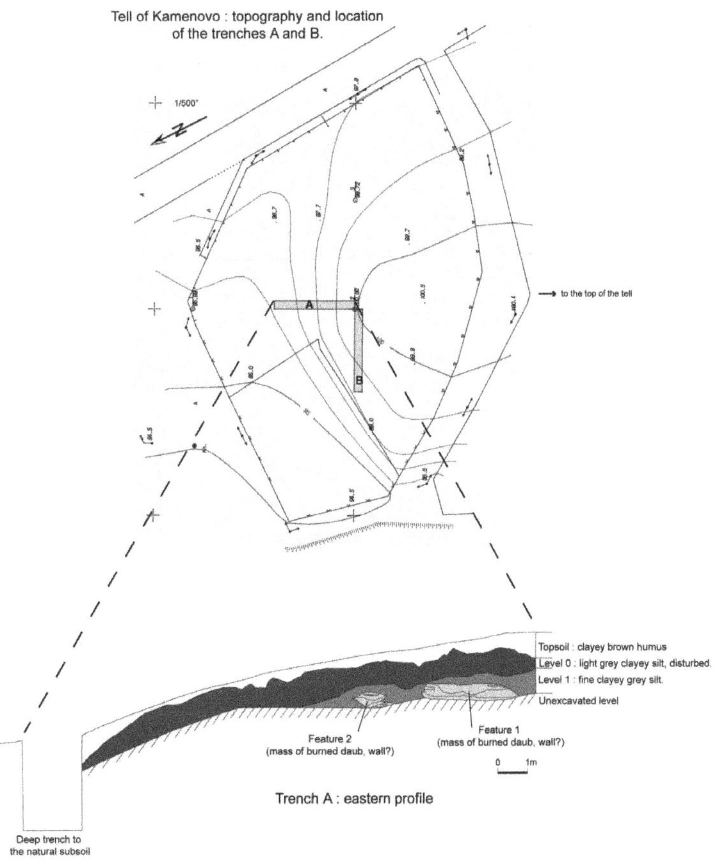

Fig. 7: Topography of the tell of Kamenovo, location of the archaeological trenches A and B (upper part), eastern stratigraphic profile of the trench A (lower part).

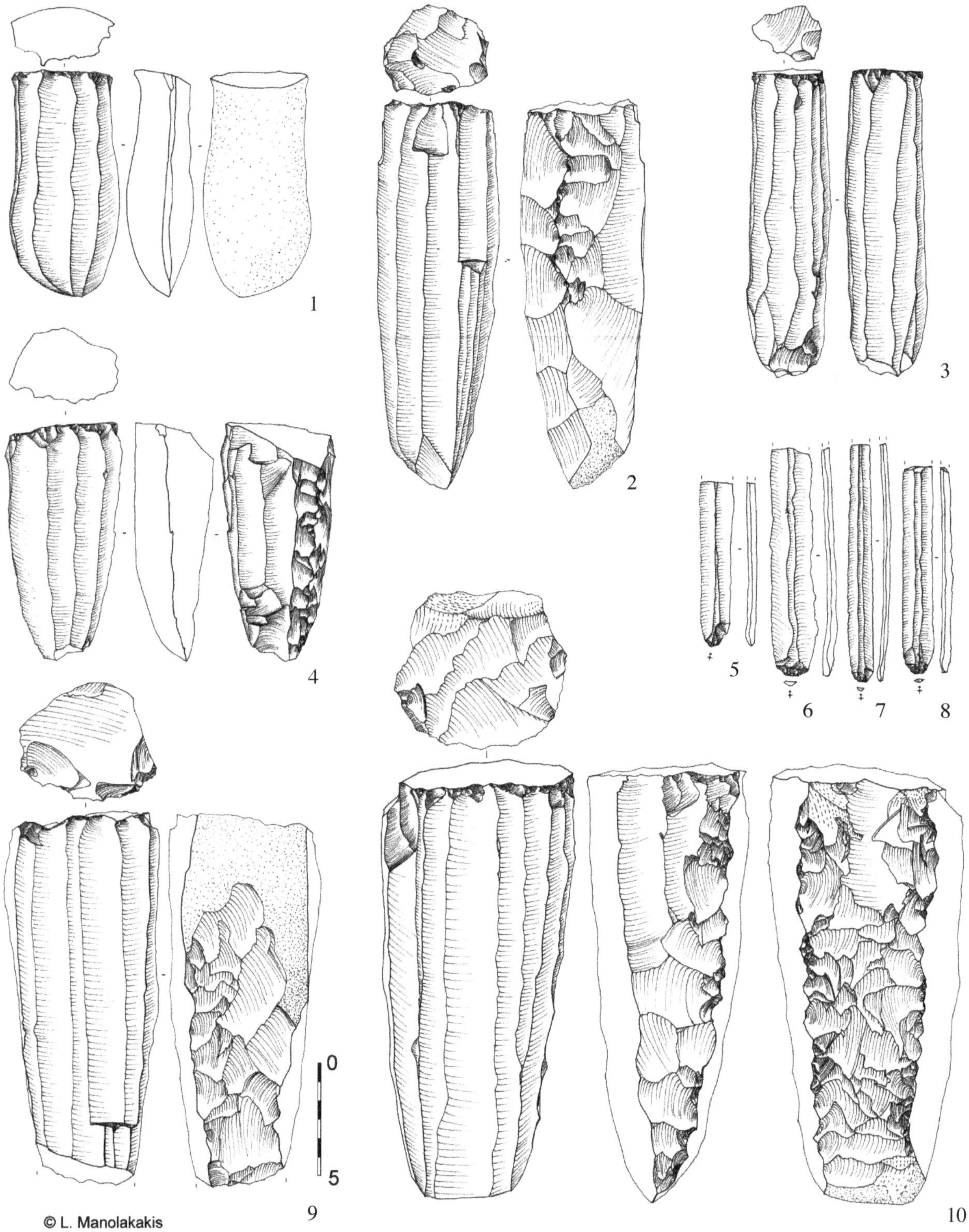

Fig. 8: Some samples of the cores from the three different knapping techniques at Kamenovo (Level 1).

Finds from level 1 are abundant and uniform. The shapes and decoration of the pottery are characteristic of Karanovo VI, late Eneolithic: lids with handles, carinated vessels, flat bases, decoration with graphite, barbotine and incisions etc. (Ivan Ivanov, in unpublished excavations report 2000). A ladle and a feeding-bottle are also part of the inventory. As well as pendants, pins, perforated discs, sawn and abraded astragali, the numerous bone and antler tools include antler punches that were probably used for indirect percussion debitage (Aline Averbouh, in unpublished excavations report 2000). There are very few hardstone finds, including a few small flakes and two small axes in undetermined green or black rock.

There are large numbers of flint artefacts, especially in the disturbed level (over 20,000 objects). In level 1, however, there is much less debitage waste and a greater quantity of full debitage blades and tools. This would appear to support the hypothesis that level 0 was a backfill.

Amongst the thousand or so lithic artefacts from level 1, punch percussion debitage is the most common (Fig. 8 n°1, 4). The cores are globular, flat or pyramidal, with a flat unique striking platform and posterior-lateral crests. The back can remain completely cortical or can be modified. The profiles of blades are curved or very curved. They are wide, with more or less regular arrises. Many blades have a flat butt, with a diffuse bulb and the overhang removed. The whole production sequence is present: from tested and discarded nodules to full debitage blades, as well as blade and flake tools. Given the presence on the site itself of nodules of adequate size, this is clear evidence for flint acquisition and debitage at the settlement.

Specialized debitage using lever pressure is also attested (Fig. 8 n°6, 9, 10), but less frequent. The lever pressure cores are long, subrectangular in section, with a flat pressure platform and straight profile. The back is modified and the sides are prepared with crests. The blades, all fragmentary, have a very small flat butt, an outstanding bulb, a carefully removed and abraded overhang, and regular, straight ridges (Fig. 8 n°10). There are large trimming and shaping-out flakes, crested blades, abandoned cores, a few fragments of large blades, tools, and axes made on cores. These finds, together with the absence of very large nodules, indicate that raw material was acquired outside the site, although debitage took place on the site.

Lastly, all the elements of the production sequence for crutch pressure are present (Fig. 8 n°2, 3), showing that this too was carried out on the site. The crutch pressure cores are short, cylindrical with peripheral debitage, or flat with a single debitage surface. The blades have the same marks as the lever pressure ones but are narrower and shorter (Fig. 8 n°5, 7, 8). This is the least abundant debitage. In fact, large nodules are not required and this sequence by crutch pressure could in theory have alternated with lever pressure on the same cores. But the shape of the cores with crutch pressure debitage is not really comparable to the cores with lever pressure debitage. Therefore, it is difficult to determine whether flint for crutch pressure was acquired on or outside the site.

The flint tools are of the kinds usually found on settlements of this period (Fig. 9): scrapers on blades (n° 13, 14) and a few on flakes (n° 15), borers and drills (n°4, 5, 6, 9) unmodified used blades, sickle blades (n° 1-3, 10-12), retouched blades (n° 8) and some retouched flakes. It should be noted that there are a considerable number of tranchet axes (Fig. 9 n° 17,18), unlike most settlements in the north-east; elsewhere, these differences in the relative frequency of tools have been interpreted in terms of preferential activities on each site.

5. HYPOTHESES ON THE MODALITIES OF ENEOLITHIC EXPLOITATION AND PROCUREMENT

Thus we are clearly in a settlement context where, as throughout the Graphite Pottery Zone, the inhabitants of Kamenovo carried out their domestic production by punch percussion. But there are two major differences:

- They located their village on a flint deposit.
- They also carried out specialized production using lever and crutch pressure.

Yet this specialized production did not exploit the deposit on which the village was located.

As for the forms of raw material exploitation, the inhabitants of Kamenovo chose to position their village on a source of good quality raw material, but they also chose the spot where the flint layer is directly accessible at the foot of the slopes, in the gully cut by the Topciiska.

Deep mine-shafts and galleries were dug elsewhere in the Graphite Pottery Zone, to extract copper ore.

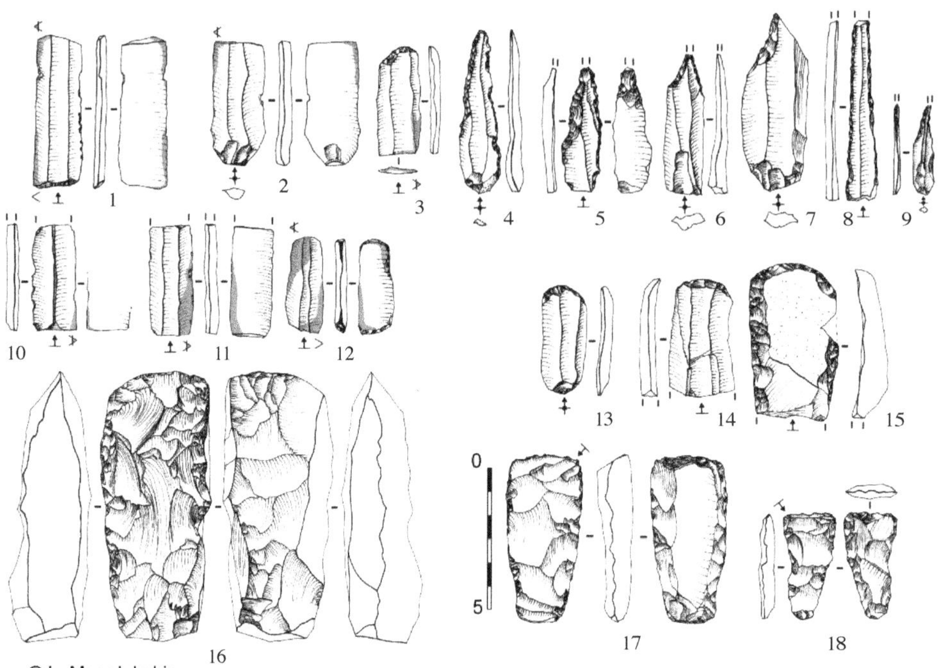

Fig. 9: Tools from Kamenovo (Level 1) : Sickles, borers, scrappers, flint axe, tranchet axes.

If one accepts the later date for the shaft at Ravno 3, one can see here, in the case of flint, that these Eneolithic populations took good advantage of what was available from a simpler form of access.

Concerning raw material procurement, there were probably two distinct modes, depending on the kind of production:

- Firstly, exploitation on the site, at Kamenovo, of flint used mainly for domestic production;
- Secondly, exploitation at a short distance, probably at Ravno 3, of raw material that was of exceptional quality, at least for its nodule size, and intended for the specialized production of very long blades.

As for debitage using a crutch, the procurement mode remains unknown: either from Kamenovo or from Ravno 3.

- Thus the two particularities of Kamenovo, specialized debitage and location on a flint source, do not seem to be connected in a cause and effect manner.
- The people did not settle there to make very long blades.

They chose to settle at a short distance from the biggest flint nodules, and this probably means that the Ravno 3 source was exclusively exploited by the specialists, as a protected area.

Everything suggests that the specialized debitage was not an intensive production. This may have been another reason to establish the settlement away from the source.

The very long blades reflect highly skilled production techniques: the blades were prestige goods of great social and cultural value in the Graphite Pottery Zone (Manolakakis 2006). Three points should be recalled here:

- These blades occurred frequently in funerary contexts, where they represented 40 to 70% of lithics.
- The longest blades were associated with the richest graves.
- The blades circulated over long distances and were present throughout the Graphite Pottery Zone, acting as a form of barrier against the obsidian which was widespread in all the neighbouring cultures.

Clearly, there is no structural link between a highly skilled, socially valued production and a mode of exploitation involving intensive production, deep mine-shafts and galleries.

One could even reverse the question: is it rather when an object is not or is no longer a prestige item that intensive or 'industrial' exploitation develops?

Laurence Manolakakis
CNRS-Protohistoire Européenne-UMR 7041
MAE, 21 allée de l'Université
92023 Nanterre, France
e-mail: laurence.manolakakis@mae.u-paris10.fr

REFERENCES

Angelov, N. 1958. Selishtna mogila pri s. Hotnica. *Studia in Honorem Acad. D. Detchev* 390-403.

Angelov, N. 1959. Zlatnoto sakrovishte ot Hotnica. *Arheologija* Sofia 1, 38-46.

Angelov, N. 1961. Rabotilnica za ploski kosteni idoli v selishtnata mogila pri s. Hotnica, Tarnovsko. *Arheologija* 3, 34-38.

Demoule, J.P. 2004. La céramique du Néolithique Récent. In R. Treuil (ed.), *Dikili Tash, village préhistorique de Macédoine orientale (Fouilles Jean Deshaye 1961-1975)*, Ecole Française d'Athènes. Paris: De Boccard, 63-270.

Hartmann, A. 1978. Ergebnisse der spektralanalytischen Untersuchung äneolithischer Goldfunde aus Bulgarien. In *Varnenskii nekropola i problemi halkolita, Symposium international, Varna, 19-23 avril 1976. Studia Praehistorica* 1-2, 27-45.

Ivanov, I.S. 1978. Les fouilles archéologiques de la nécropole de Varna (1972-1975). *Studia Praehistorica* 1-2, 13-26.

Lichardus, J., Lichardus-Itten, M., Bailloud, G. and Cauvin, J. 1985. *La protohistoire de l'Europe. Néolithique et Chalcolithique.* Paris : PUF, Nouvelle Clio 1 bis.

Manolakakis, L. 2005. *Les industries lithiques énéolithiques de Bulgarie.* VML Internationale Archäologie 88.

Manolakakis, L. 2006. Les très longues lames de Varna, quelle fonction? In J. Vaquer et J. Guilaine, *La fin de l'Age de Pierre en Europe du sud,* Actes de la table ronde de l'EHESS. Editions des Archives d'Ecologie Préhistoriques, Toulouse, 2006. 5-23.

Manolakakis, L. and Ivanov, I.S. (eds.) 1995-1999. *Les ateliers de taille du silex de Bulgarie du Nord-Est.* Unpublished excavation Reports. Mission franco-bulgare, Ministère des Affaires Etrangères.

Manolakakis, L. and Ivanov, I.S. (eds.) 2000. *Les fouilles du tell de Kamenovo, habitat et atelier de taille chalcolithique.* Unpublished excavation Report. Mission franco-bulgare, Ministère des Affaires Etrangères.

Natchev, I., Kovnurko, G. and Kantchev, K. 1981. Kremacnite skali v Balgarija i tjahnata esploatacija, *Interdisciplinarii Izsledvanija.* Sofia: Arheologiceski Institut i Muzeï na BAN, VII-VIII, 41-58.

Natchev, I. and Kantchev, K. 1984. Aptian and quaternary flint in Noerth-East Balgaria, *Report of the III° seminar in Petroarchaeology, Plovdiv*, 65-82.

Tchernih, E.N. 1978. Aibunar, a Balkan copper mine of the fourth millenium BC. *Proceedings of the Prehistoric Society* 44, 203-217.

Tchernih, E.N. 1978. O jugo-vostotchnoi zone Balkano-Karpatskoï metallurgitchesko provincii epohi eneolita. In '*Varnenskii nekropola i problemi halkolita*', *Symposium international, Varna,* 19-23 avril 1976. *Studia Praehistorica,* 1-2, 170-181.

Todorova, H. 1978. *The Eneolithic in Bulgaria.* BAR, int. ser. 49. Oxford: BAR.

FLINT MINING IN EARLY NEOLITHIC IBERIA: A PRELIMINARY REPORT ON 'CASA MONTERO' (MADRID, SPAIN)

Marta Capote, Nuria Castañeda, Susana Consuegra, Cristina Criado, Pedro Díaz-del-Río

Abstract: The Early Neolithic (c. 5400-5100 cal BC) flint mine of Casa Montero was located in 2003 and partially excavated throughout three field seasons. This paper describes the excavation strategies that have been applied –both random selection of shafts and aligned systematic sampling–, and some of the preliminary results obtained from excavation and post-excavation analyses. We conclude that the dimensions of the site and the characteristics of the shafts suggest short term seasonal expeditions of small mining teams. This mining process produced use and probably exchange value, but also a durable monumental landscape, erected as a result of a cumulative materialization of social labour.

Keywords: Flint mine, Neolithic, Iberia, operative chain, shafts, labour, monuments, landscape.

INTRODUCTION

Until the nineties most Iberian prehistoric flint mines had been located in southern Spain through surface collections. Of these, barely one had been explored (Ramos Millán 1997; Ramos Millán *et al.* 1997). All lacked absolute chronologies, and from what we now know, many were actually mined in modern times for military purposes. In a similar way, Neolithic studies have only recently committed to long-term regional research in areas other than the well-known Levantine coast.

This is particularly the case for the Central Meseta, until fairly recently an area almost devoid of Early Neolithic evidence. Our present knowledge relies on a few recently excavated sites, most known through preliminary reports and only one completely published. There are two different categories of sites: caves and open-air settlements (Bueno *et al.* 2002; Díaz-del-Río and Consuegra 1999; Estremera 2003; Kunst and Rojo 1999; Rojo and Kunst 1999; Rodríguez 2006). While the nature of the caves and the limitations of the small areas where digging took place do not allow a precise functional interpretation of the earliest Neolithic layers, open-air settlements are mostly made up of small unimpressive clusters of pit-features. One of their shared features is the limited amount of portable remains recovered in early Neolithic layers. This is especially true when quantifying flint tools, which seem to be extremely scarce, compared both to other contemporary evidence elsewhere, or to the subsequent regional Copper Age settlements (Díaz-del-Río 2004).

Under these circumstances, the discovery, in 2003, of the Early Neolithic flint mine of Casa Montero in the centre of Iberia was a striking novelty on account of its location, early chronology, density of shafts, and the historical context in which flint mining developed. The present paper is an overview of the evidence recovered after three years of fieldwork, and a preliminary report on post-excavation analyses in progress.

Fig. 1: Location of Casa Montero in Iberia and in relation to the Jarama basin.

1. THE SITE AND THE DIGGING CAMPAIGNS

The flint mine of Casa Montero was discovered as a result of the Archaeological Impact Assessment of Madrid's M-50 highway belt. The site is situated in the south-east of Spain's expanding capital city, in the centre of Iberia (Fig. 1). Located on a river bluff, the position dominates one of the main regional river basins, the Jarama valley, where some scattered Neolithic sites have been known to exist (Mercader *et al* 1989; Díaz-del-Río and Consuegra 1999). The results from three excavation seasons combined with an intensive surface survey with systematic sub-plowzone control pits suggest that the prehistoric mine may roughly cover eight hectares.

Four chronological phases were documented in the 4 hectare open-area excavations undertaken at the site. These offer evidence of Middle Pleistocene and Bronze Age occupation, and Neolithic and Contemporary flint mining exploitations. The oldest phase was located in the north of the excavation, characterized by an open-air Pleistocene stratigraphic sequence.

The Neolithic evidence includes 3824 small vertical shafts, one meter wide on average, and of up to nine meters deep, dependent on the variable depth and quality of the flint seams. The lithic record is exceptional, both in quantity and quality, and includes all phases of the operative chain. Radiocarbon dating of two charcoal samples have afforded early Neolithic dates that suggest that the flint mine was exploited throughout the second half of the VI millennium cal BC (Díaz-del-Río *et al.* 2006). To date, Casa Montero is, after Defensola, (Galiberti *et al.* 1997; 2001), the second oldest Neolithic flint mine in Europe (Fig. 2).

Bronze Age pit structures are found in small percentages, both in between the cluster of Neolithic shafts and in its contiguous area. None can be considered to be mining

Ref. Lab.	Date BP	Dated material	Cal 2 sig	Provenience	Meth
Beta 206512	6410 +/- 40	Carbon *Quercus ilex/ Q. coccifera*	5480-5320 BC	Shaft ue. 2384 Layer ue. 2382	AMS
Beta 206513	6270 +/- 40	Carbon *Quercus ilex/ Q. coccifera*	5330-5070 BC	Shaft ue. 2701 Layer ue. 2229	AMS

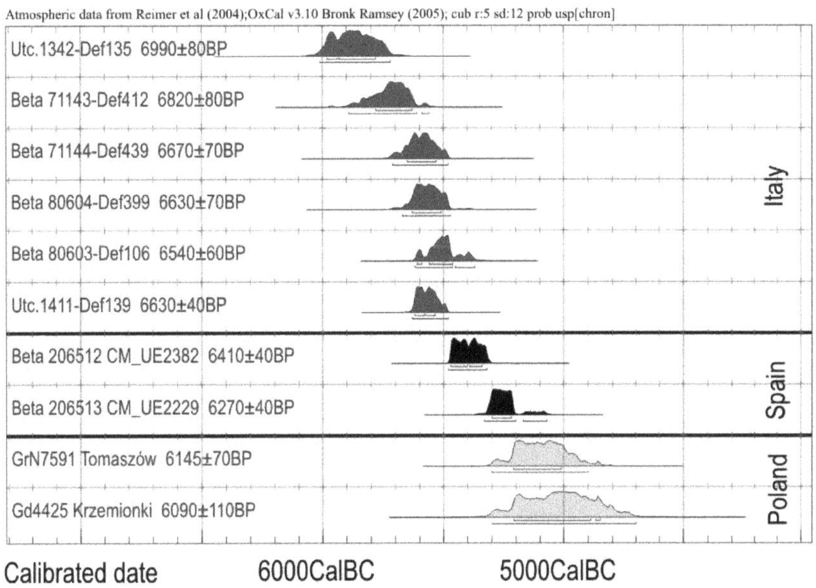

Fig. 2: Calibrated radiocarbon dates of the earliest Neolithic flint mines in Europe: Defensola, Casa Montero, Krzemionki and Tomaszów.

shafts, but they share characteristics with all regional BA sites: the pits are mostly related to storage and waste disposal. These shallow pits include a few crude throwaway flint tools, potsherds, charcoal, some faunal remains, and occasional individual burials. Many BA sites seem systematically to reoccupy certain previously significant sites, like Copper Age enclosures (Díaz-del-Río 2004) or this Neolithic flint mine. The distribution of features also shows a pattern: a few structures on the previous site, while most cluster around in its immediate vicinity. Considering the extremely irregular topography of the mine by the end of the Neolithic, and the limited use of flint during the Bronze

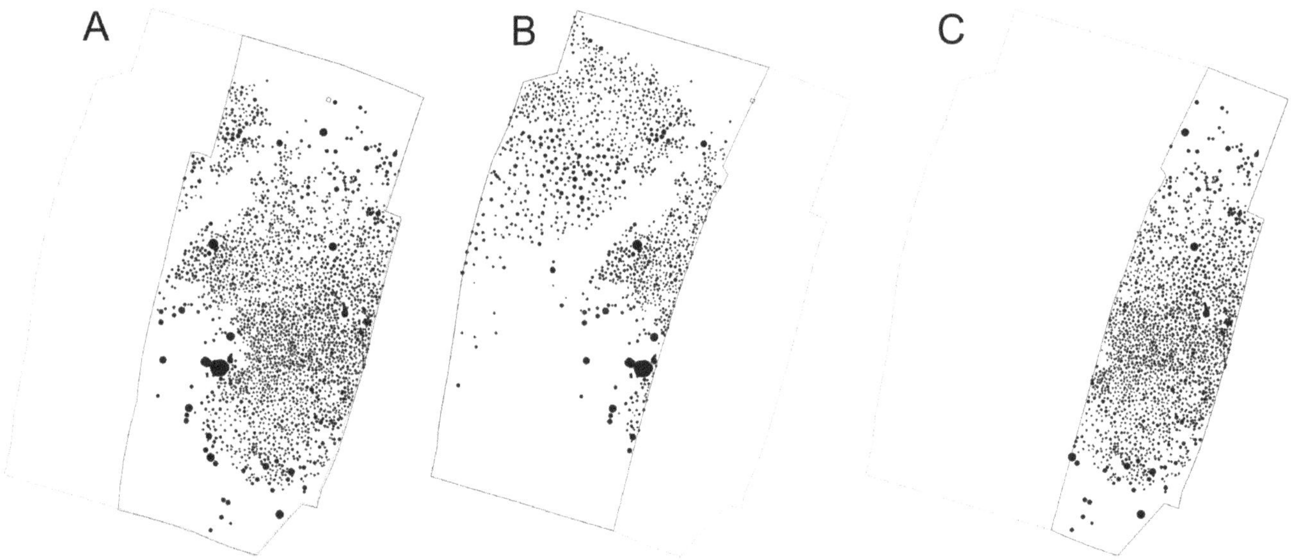

Fig. 3: (A) Area documented during the first field season. (B) Area documented and sampled during the second and third field seasons. (C) Documented area that has been preserved and protected.

Age, it is functionally unclear why scattered BA groups would occupy this particular spot. Nevertheless, the presence of two individual burials suggests that at least some ritual behaviour may have been involved. Whatever those groups thought about previous flint mines or enclosures, the pattern seems to suggest a strong interest in claiming a connection with their -real or fictitious- ancestors, wherever they had left visible earthworks.

Finally, the surroundings of Casa Montero are known today as 'Cerro de las Canteras' (Quarry Hill). Both the Neolithic site and its surroundings were mined for flint to be used in threshing machines and possibly for gunflint. The historical record suggests that they may have been under exploitation until the mid-Nineteenth century (Consuegra *et al.* 2004; Castañeda and Criado, 2006; Prado 1864).

The site of Casa Montero was excavated during three field seasons, from September 2003 to July 2006, under different circumstances and with varying research aims (Fig. 3). The first season (September 2003-March 2004) involved the excavation of 123 shafts, randomly selected from the approximately 2500 shafts mapped in 2.4 hectares of open-area excavation. The objective was a first assessment of the variability of prehistoric flint extracting methods, the geological structure of the site, and the absolute and relative dating of the site through the available evidence: charcoal remains and pottery fragments (Bustillo and Pérez-Jiménez, 2005; Capote *et al.* 2006; Castañeda *et al.* 2008; Castañeda and Criado 2006; Consuegra *et al.* 2004; 2005; Díaz-del-Río *et al.* 2006; Díaz-del-Río *et al.* 2008; Pérez-Jiménez *et al.* 2005).

That Casa Montero was the first Neolithic flint mine discovered and extensively dug in Iberia gave regional Heritage Managers some arguments to push for a modification of the highway belt some 60 meters to the west from its original track, saving from destruction and permanently protecting the main concentration of prehistoric shafts. As a result, during the second season (August 2004-February 2005), the open-area excavation was extended 1.8 hectare further west, and more than 500 new shafts were mapped. This excavation led to reconsideration of our research objectives, as key matters such as mining strategies, contemporaneousness of extractions or other issues could only be potentially answered by recovering information from clusters of shafts.

Consequently, during the third season (September 2005-July 2006) the excavation strategy shifted from a focus on individual mining shafts to groups of shafts. In order to control potential horizontal variations in mining strategies or chronology, we established an aligned systematic sampling. The approach involved the complete excavation of all the shafts included in each one of the seventeen grid squares of ten by ten meters systematically distributed throughout the area that would ultimately disappear under the highway (Fig. 4).

For safety reasons, this sampling was accompanied by a different excavation technique. All selected shafts were dug down to two meters deep, followed by the complete extraction of the remaining geological layers with a backhoe. The grid squares were once again topographically located and the excavation process repeated four times until most of the deepest shafts had been completely dug out. Of all these, only three shafts were deeper than 10 meters and not completely excavated. As a result, we obtained both detailed information on shafts and a precise three dimensional view of the geological structure of the complete excavation area. This new excavation technique allowed for the recording of several narrow passages that connected groups of shafts, a result that would have been impossible if we had followed the previous single shaft excavation procedure. Neolithic miners would have crawled through these passages following some horizontally disposed flint seams.

2. GEOLOGICAL CONTEXT

Casa Montero is located in sedimentary rocks from the Intermediate Unit of the Miocene in the Madrid basin (Brell *et al.* 1985; Junco and Calvo 1983). The stratigraphic column is composed of beds of clay, dolomite and silica rocks. Deep sections show the existence of four major silicification episodes, each consisting of one or more silica levels. The three upper levels are composed of opal and opaline chert, and were formed by silicification of Magnesian smectites (Pérez-Jiménez *et al.* 2005; Bustillo and Pérez-Jiménez 2005) (Fig. 5).

Silica rocks from Casa Montero form nodules arranged in discontinuous beds that may have some lateral continuity. These beds have been classified in four episodes of silicification (Bustillo and Pérez-Jiménez 2005); they appear deformed as a result of collapses of the underneath evaporitic episodes. This deformation produced a depression in which most of the shafts are concentrated, and

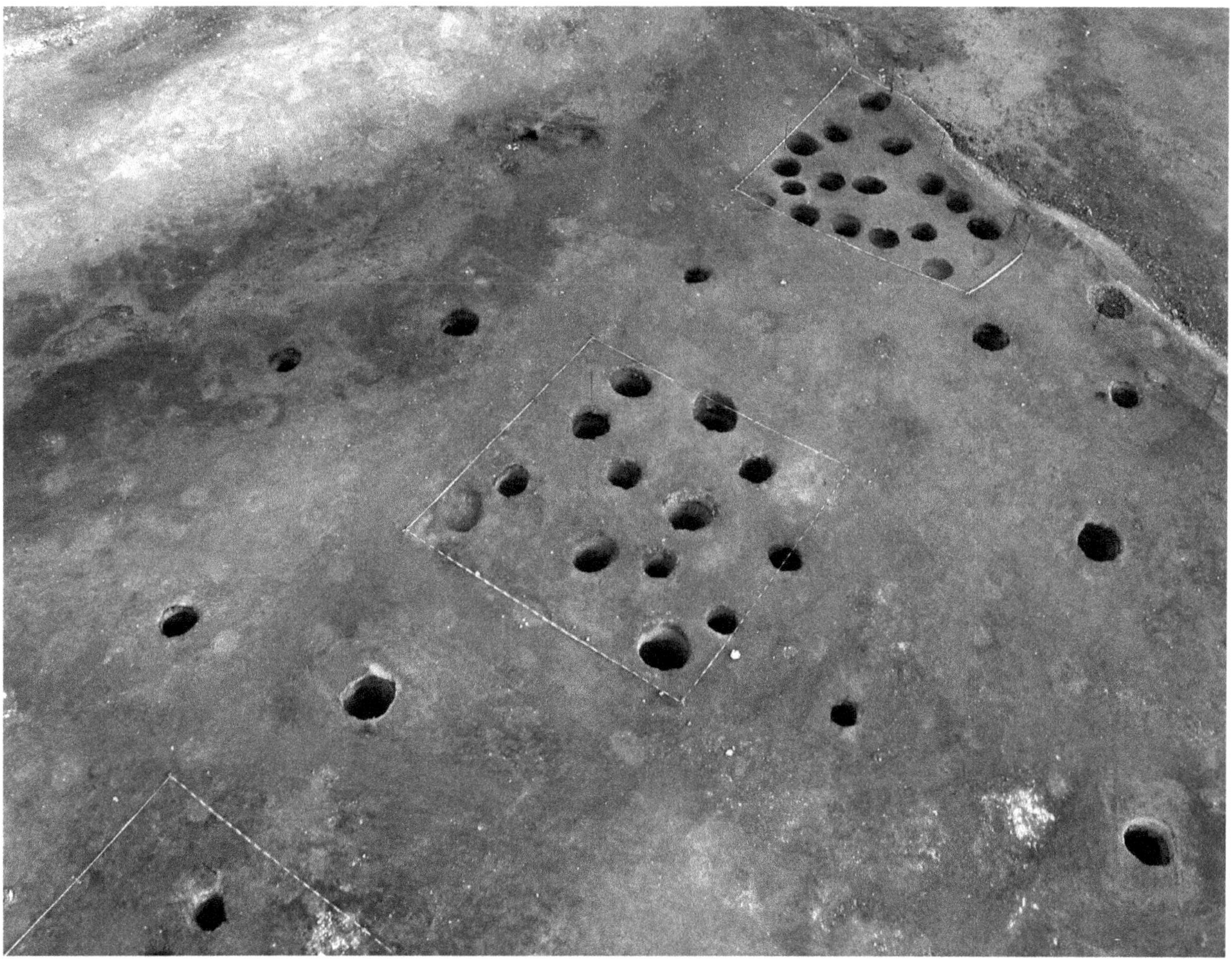

Fig. 4: Aerial view of three of the 10x10 meter areas during their excavation.

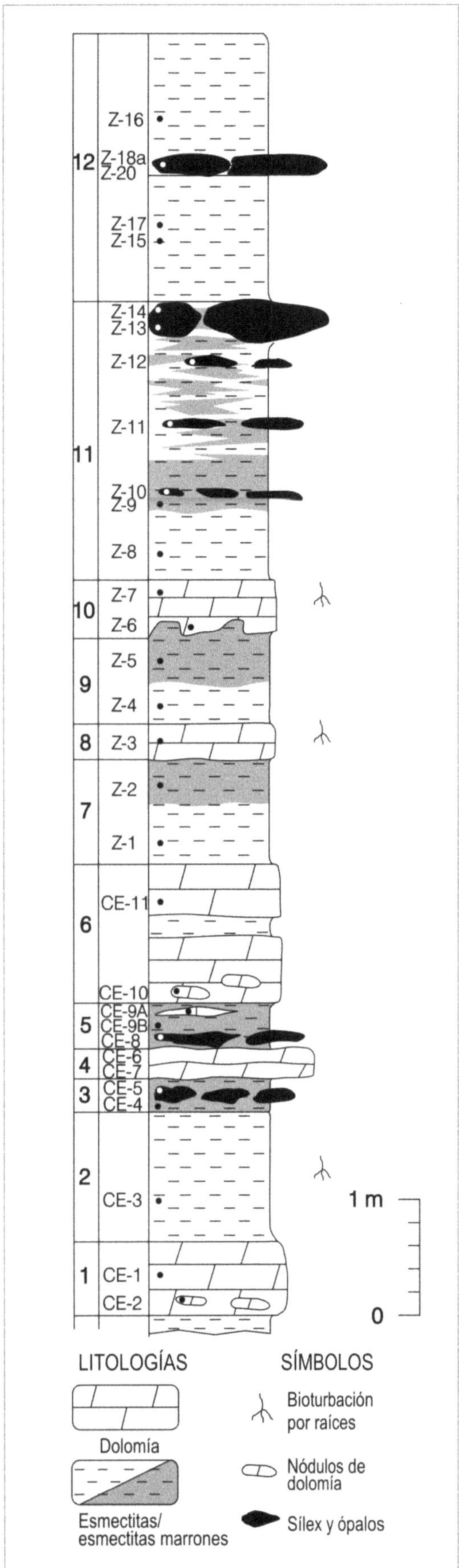

Fig. 5: Stratigraphical column documented at Casa Montero.

may be particularly related to the horizontal depth variability of shafts throughout the site: as a general pattern, shafts are deeper in the central area of the excavation.
The selection of this particular mining area was no coincidence. The regional geology of Madrid is known for its abundance in siliceous rocks; that is, the profusion of flint at this site cannot be considered a uniquely distinctive feature. Although we cannot rule out the possibility of other regional flint mines, Casa Montero's flint has a particular genesis: it was formed from smectites and later underwent an aging process. This involved a re-crystallization of opal to form quartz in the inner part of nodules, developing a fine-grained core. As a result, there are frequent nodules with opaline outer parts and microcrystalline quartz inner parts (Bustillo and Pérez-Jiménez 2005). This process gave the flint particularly suitable knapping qualities (Pérez-Jiménez *et al* 2005; Bustillo and Pérez-Jiménez, 2005; Bustillo *et al.* 2008).

3. RAW MATERIALS

Casa Montero's lithic remains have been classified in two groups of raw materials. First, rocks which were necessarily obtained off-site, and used in the process of extraction, quartering, and reduction of flint nodules. Secondly, silica rocks, which were the main object of the mining activity.

Off-site rocks are quartzite, quartz, and sandstone pebbles, most probably obtained from the nearby Jarama river bank, less than a kilometre away. Such rocks constitute a minimal part of the overall assemblage.

The second group is composed of siliceous rocks: opals and opaline cherts. Four silicification episodes have been documented. Neolithic shafts cut through the three upper episodes, all opaline. The main exploited levels are the second and fourth, while the third is only exploited whenever it has good knapping qualities (Fig. 5).

We undertook a macroscopic and petrographic characterization of the raw materials represented in the archaeological record, in order to analyze in depth the cultural and/ or physical characteristics that had determined either its selection or refusal. In order to carry out this characterization, a sample of 43 lithic remains was selected out of a population of 5043 artefacts obtained from the first field season. The macroscopic characterization has allowed us to distinguish seven types of raw materials. Subsequently, the petrographic study has defined the properties of the opaline cherts.

Refits and chaîne opératoire analyses have provided evidence indicating that when nodules have an opaline outer part, this material is worked as a second phase of removal of cortex, whereas the raw material chosen for lithic exploitation is chert from the inner part of the nodules. Precisely this inner part is the one reduced in the phases of configuration and production. The reason for this selection is knapping quality, since opal is fragile and, consequently, less suitable than chert as a blank. Moreover, the frailty of opals cause fractures, faults and errors in the process of flaking. As a result, a large quantity of this raw material inevitably has to be discarded. On the other hand, the heterogeneity of nodules always involves additional difficulty for flint-knappers, as the response of each material to the force of the blow will be unexpected and consequently difficult to control (Castañeda *et al.* 2008). (Fig. 6)

Fig. 6: Flint refits from Casa Montero

4. THE MINE

The archaeological record of Casa Montero is the result of sequential short term activities which took place over a relatively short period of time. In order to tackle this typical problem of time scales, the information has been contextualized and sorted out distinguishing different working processes. The labour process involved at least the following succession of actions, each one of them preceded by specific decision-making in respect to parameters such as space, quality and quantity: excavation of the shafts, extraction of raw materials, transformation of raw materials, and waste management. Each action has produced a particular kind of archaeological record.

4.1. Shaft excavation

In order to analyze this action, we rely on the following evidence. First, the mining structures themselves and their spatial distribution. Second, the mining tools employed in the process of shaft digging and recovered in the archaeological record. Finally, the marks left by mining tools on shaft walls and the traces of preparation of structures to facilitate both access to shafts and evacuation of mine spoils.

Mining shafts offer little variability in respect to size and shape. They are mainly simple cylinder-like structures on average one meter wide and up to nine or ten meters deep. Their infillings show little differentiation, and few archaeological remains other than an impressive amount of flint. However, one of the sampling units dug during the last field season revealed a cluster of five shafts with small connecting galleries (Fig. 7). These were the result of side workings dug to exploit the one meter deep seam. This seam may have been abandoned when nearly exhausted or when the costs and efforts to continue it exceeded those of opening a new shaft. Occasionally, several of these small connecting galleries are documented at different depths in a single shaft. Although the system is more complex and labour intensive than the straightforward excavation of vertical structures, it is still extremely simple, given the total scale of the mining activity.

Fig. 7: Side workings documented at one of the sampled areas.

As has been demonstrated for other Iberian mines (Camprubí *et al.* 2003), Neolithic miners certainly had a good knowledge of the geological structure of the area. Test pits at Casa Montero are an excellent testimony to this knowledge (Consuegra *et al.* 2004). The Neolithic miners stopped excavating the shallow pits whenever they found green clay levels that are stratigraphically located under the opaline episodes. Furthermore, the depths of the shafts adapt to deformations resulting from siliceous episodes. This geological know-how would have been the result of a transmission of local mining knowledge from generation to generation: extraction and debitage methods, best conditions for lateral excavations, safety measures, and so forth. Shafts were dug close together, none superimposed on another, and with just enough distance so as to walk between them while avoiding wall collapses. These more than 4000 Neolithic shafts are probably the result of several centuries of mining. The total mining intensity would have been about 13 shafts per year, considering a time-span of 300 years for the whole period of activity.

Antler and bone tools for shaft digging are commonly found in other European mines such as Jablines (Bostyn and Lanchon 1992; 1997), Grimes Graves (Holgate 1995; Topping 1997; Barber *et al.* 1999), Zele (Lech and Lech 1984; 1997) or Krzemionki (Migal and Salacinski 1997). These tools may be picks, chisels, shovels or even rakes. On the contrary, most of the tools related to mining activities recovered so far from Casa Montero are made out of stone. They are picks, maces or wedges, as is the case for other mines like Ryckholt (Felder 1997) or Defensola (Galiberti *et al.* 1997; 2001; di Lernia *et al.* 1995). The very rare antler fragments recovered during the excavation seem to relate more to small bone utensils such as awls or pressure flakers than to picks (Yravedra *et al.* 2008). Nevertheless, the information recovered from several casts of tool marks on shaft walls suggests the use of bone implements in at least parts of the mining process. Both the analyses of these casts and of possible use wear on the antler points will be decisive for their interpretation.

Lithic mining tools have been classified in two groups, according to the kind of activity they permitted: raw material procurement tools -composed of shaft digging tools (picks, maces or wedges)- and maintenance tools -mostly scrapers, denticulates and unmodified flakes- (Consuegra *et al.* 2004; 2005).

Two types can be identified among the raw material procurement tools: striking tools and incisive tools. The first group includes maces and big hammerstones; the second picks and wedges.

Striking tools were heavy, with rounded forms. They were manufactured using raw materials obtained off-site, like quartzite from the nearby terraces of the Jarama River. An example of this kind of tool is a mace manufactured from a quartzite pebble, with prismatic shape and a quadrangular and homogeneous section (Fig. 8). It measures 137x81x78 mm and weights 1,400 grams, presenting two use areas on both ends, with traces of an intensive percussion that provoked numerous accidental extractions. The mace has a medial zone with traces of crushing and abrasion, a consequence of the elimination of protuberances and conditioning for hafting. Maces may have been multifunctional, used for both the excavation of shafts and the fragmentation of big nodules (Consuegra *et al.* 2005).

Fig. 8: An example of a flint pick and a quartzite mace from Casa Montero.

Incisive tools, on the other hand, were usually selected from discarded fragments that resulted from in situ knapping activities. These fragments were chosen because of their size, weight and pointed or angular morphology. Configuration is practically nonexistent in most of these objects. When it does, it is simply performed in order to facilitate hafting. They all show evident traces of percussion and use.

Ten flint picks have been identified so far (Fig. 8), one of them made out of silicified clay. The average dimensions of these items are 103x53x27 mm. Their width and thickness are normally more homogeneous than their length. Working areas are usually very homogeneous, with a triangular or more rarely a quadrangular or trapezoidal section; they usually are 15 mm thick. Most of them show longitudinal and oblique striations on their active areas, and abrasion on ridges. On most of them, microflakes and traces of abrasion have been observed in their medial and proximal ends. The light weight of these picks suggests that they may have not been used by themselves, but instead, employed as chisels in combination with maces (Consuegra *et al.* 2005).

We have classified denticulates, scrapers, endscrapers and burins as maintenance tools. They were all expediently manufactured from discarded flint fragments, and probably used in activities such as the manufacture and repair of hafts, ladders or ropes. Non-modified flakes that resulted from knapping activities would also have been employed for these maintenance activities. These would not have required any further modification, as the presence of flakes with macroscopic use-wear traces -especially microflakes- seems to suggest.

Digging tools have occasionally left marks on shaft walls. Most of these are longitudinal marks with V or occasionally U sections. They appear in groups of parallel marks with vertical or oblique disposition, although exceptionally some shallow and circular marks have been documented. Only occasionally do some shaft walls show certain modifications such as small depressions to facilitate access, or pairs of small opposite holes with circular section to support some kind of pulleys. As the mean diameter of shafts does not allow for more than one miner at a time, any shaft more than 2 meters deep would require a minimum of two individuals to work it. The assistant stationed above would help with the process of waste disposal, hauling up the raw material, and handing down tools to the miner below.

4.2. Raw material extraction

This stage comprehends all the tasks performed in order to extract nodules from flint seams, quarter them, and raise the nodules and fragments to the surface. Although shafts do cut through flint seams, miners did not always exhaust these, so some seams are still visible in the shaft walls. They occasionally show evidence of intentional fractures, or the voids resulting from the withdrawal of whole nodules.

Resistance to fracture is greater when nodules are inserted in their matrix, and knapping becomes increasingly difficult inside narrow shafts that do not allow the miner to choose an adequate angle to strike. Depending on the size of the nodule, and its position inside the shaft, nodules would have been extracted whole or quartered and extracted in large flakes. Those nodules small enough to be manipulated would have been directly removed from flint seams, while the bigger ones would have been fractured with the aid of big hammerstones, maces and wedges. Accordingly, there are few examples of exploitation of complete nodules among the lithic production of Casa Montero. As a rule, large flakes are extracted from nodules because they make more suitable core blanks (Consuegra *et al.* 2005).

Tools used in raw material extraction can be divided into two groups: incisive implements such as wedges, and percussion implements such as big hammerstones.

Some wedges have occasionally been recovered at the site. These implements were either manufactured from discarded pieces resulting from flint knapping, or from fragments of broken quarzite hammerstones, selecting those with a particular morphology, size and weight, without any specific configuration. Consequently, these wedges are improvised implements, as happens with all the tools so far documented at Casa Montero. This kind of incisive tool shows a thin distal end that increases in thickness towards its medial part and proximal end. Its working area is an edge with highly variable length. This working edge has macroscopic marks on the edge, and striations parallel to the main axis of the artefact. The proximal end of these tools also display macroscopic use wear traces, such as crushing and accidental extractions. These traces are percussion marks resulting from the use of wedges as intermediate elements. The edge of the tool would have been inserted into cracks or crevices in shaft walls, and the proximal end would have been struck with a mace in order to extract the nodules.

The frequent difficulties occurring while quartering and extracting nodules required the use of heavy hammerstones and, occasionally, the use of maces for shaft digging. These big hammerstones have been recovered either complete or –more often– as fragments. They are always manufactured from quartzite, with a spherical or elliptical morphology, an approximate length of 163 mm and an average weight of 1200 gr. They show many traces of hard percussion producing extensive accidental flakes and/or longitudinal fractures.

4.3. Raw material transformation.

At the Casa Montero flint mine, the raw material was both procured and processed. This has resulted in a massive amount of recovered archaeological remains: lithic materials recovered from the three digging seasons weigh 65 T, and amount to more than one and a half million pieces. This abundance of material creates extreme difficulties for its efficient management, processing, analysis and comprehensive publishing. On the other hand, it offers the advantage of allowing the complete study and careful reconstruction of the *chaîne opératoire* (Consuegra *et al.* 2004; Castañeda and Criado 2006; Castañeda *et al.* 2008).

A preliminary analysis of lithic remains from the first field season led us to understand the main aim of flint exploitation: the production of blades. There is also a secondary, although important, flake production, identified with its own independent chaîne opératoire. This kind of production is peculiar to Casa Montero, especially when compared to other European Neolithic flint mines that mostly aimed at the production of axe rough-outs, such as Grimes Graves (Barber *et al.* 1999), Jablines (Bostyn and Lanchon 1997) or Defensola (Galiberti *et al.* 1997). Examples of Neolithic flint mines aiming at blade production, such as Tomaszów (Schild 1995), are somehow less common in Western Europe.

Evidence from Neolithic contexts suggests that blades played both a functional and ritual role. Blades have been documented in early Neolithic individual burials from Central Iberia, such as La Lámpara (Rojo and Kunst 1999), although they later become a generalized funerary offering in megaliths (e.g. Rojo *et al.* 2005; Delibes *et al.* 1993). Not infrequently, blades were specifically manufactured for burials, showing no evidence of wear for use.

Blade production is documented at the mine through blade cores, crested blades, discarded blades and knapping accidents. Blade products of good quality do not appear, mainly because they were removed, probably along with some configured blade cores. Neither used nor retouched blades have been recovered at the mine, that is, tool configuration and use are not represented in the blade *chaîne opératoire*.

The exceptional lithic record of Casa Montero allows for the recognition of different blade production schemes. The choice of one or another reduction strategy depends on the nature of the blank –a complete nodule or a large flake- and on the blade needed. The lithic record analyzed from the first field season suggests that the preferred raw material for blade production was flint rather than opal. As documented through several refits, this preference required the elimination of all opal parts previous to the configuration of the flaking surface (Consuegra *et al.* 2004; Castañeda *et al.* 2008).

When the blank is a whole nodule, the systems developed are mainly prismatic, beginning with the preparation of a crested blade. Flake blanks are more suitable and versatile than entire nodules. These large flakes allow for different reduction strategies. So far, three reduction schemes for blade production have been documented, two in volume and one in surface. Each will produce a particular kind of blade: either more elongated and thin or shorter and thicker (Castañeda and Criado 2006; Díaz-del-Río *et al.* 2006).

Flake production has been identified thanks to the presence of predetermined and predetermining flakes and their corresponding cores. Flakes are obtained through bifacial exploitation systems, either hierarchical or non-hierarchical: both discoidal and levallois schemes have been identified (Consuegra *et al.* 2004).

Hammerstones used for knapping activities were collected and selected among materials from the Jarama river bank. They are quartzite, quartz and sandstone pebbles that can be sorted in two different volumetric groups: medium and small, fitting an average hand.

4.4. Waste Management

The soil extracted during the excavation of shafts was either dumped into other nearby shafts along with the remaining waste from flaking, or left aside and finally dumped back in. This of course depends on whether miners opened more than one shaft at a time. It seems reasonable to suggest that the second option was preferred: we lack evidence for natural sediments at the base of any excavated shaft, and layers that can be interpreted as wall collapses are rare. In any case, it seems that shafts were filled almost immediately after being excavated. Shaft in-fills show a repetitive pattern where density of lithic finds increases with depth. This behaviour would keep working areas free from spoils and flaking residues, and, given the proximity of shafts, the systematic backfilling may have been the most practical safety measure.

5. EVIDENCE OF OTHER ACTIVITIES

Miners at Casa Montero understandably performed other activities. Among the archaeological evidence recovered are pottery fragments, bone industry, faunal remains and ochre fragments.

The few fragments of pottery recovered at Casa Montero are in form and decoration characteristic of central Iberia's Early Neolithic, similar -if not identical- to those recovered at Cueva de la Vaquera (Estremera 2003; Consuegra *et al.* 2004). Frequently they are fragments of impressed wares (Fig. 9), occasionally decorated with red dip-coatings, and their shapes suggest mostly middle-sized containers. Pottery will be analyzed for phytoliths, and thin-sections will be obtained in order to define the potential technological and/or compositional variability between wares. These may inform us about functional differences between wares, or differences in the provenance of pottery productions, perhaps related to the social conditions of accessibility to the mine.

Fig. 9: Two impressed ware fragments recovered at Casa Montero.

On the other hand, there is a small assemblage of bone tools, most of them corresponding to bone and antler shafts, and finished rings that were produced and abandoned on site (Yravedra *et al.* 2008). This evidence suggests that the manufacture of bone and antler rings was somehow required while labouring at the mine. Although the function of these rings and their relation with flint mining are yet to be found, it seems that these artefacts -typical of Iberian Levantine Early Neolithic sites (Pascual Benito 1996)- should be considered rather as functional than ornamental, as previously thought. Whatever their function, they must relate to an activity performed in both mining and non-mining contexts, and they are one of the best chronological markers of Casa Montero.

Finally, ochre impregnated on lithic elements and bones, has been documented in considerable quantity, as well as a large 2 kg portion of pigment. Ochre has usually been related to both ritual and functional contexts: it was used in funerary monuments, to process different raw materials, and in the production of elements such as hafts. Nevertheless, there is so far no unambiguous evidence for Neolithic ritual activities at Casa Montero, as opposed to other Iberian (Bosch and Strada 1994) or European mines (e.g. Barber *et al.* 1999; Topping 1997). However, mining itself may have been a ritualized practice.

6. CONCLUDING REMARKS

Some preliminary conclusions can be drawn from Casa Montero, especially when compared to European Neolithic flint mines in general.

The first observation bears on the scale of mining activity. This Iberian mine is noticeably small in extension, compared to many other western European flint mines. This is not so strange, if we consider the most probable low population densities of the Madrid region during the late Sixth millennium BC, and the 300 year life-span of the mining activity suggested by radiocarbon dates (see Fig. 2). These similarities are less obvious when comparing Casa Montero with the flint mines of Defensola (Galiberti *et al.* 1997; 2001), its contemporary central Mediterranean counterpart. Although they are both associated with the first introduction of a production economy in their respective regions, differences in the size and complexity of labour and mining techniques seem obvious.

The shafts are also considerably smaller than in most flint mines, with the 117 cm mean diameter of shafts at Casa Montero[1] (Fig. 10). They allow for no more than one miner working inside at a time, all of which suggests that the minimum Neolithic working team may not have been more than two individuals. However, we should consider the possibility of some labour coordination throughout the whole mining and knapping procedure, as seems to be the case in most known ethnographic cases (Burton 1984; Stout 2002). Whatever the size of the mining teams, the width of the shafts seems to result from the combination of a limited availability of labour force and practical reasons related to the geological structure of the area, where collapses are common when clays are exposed.

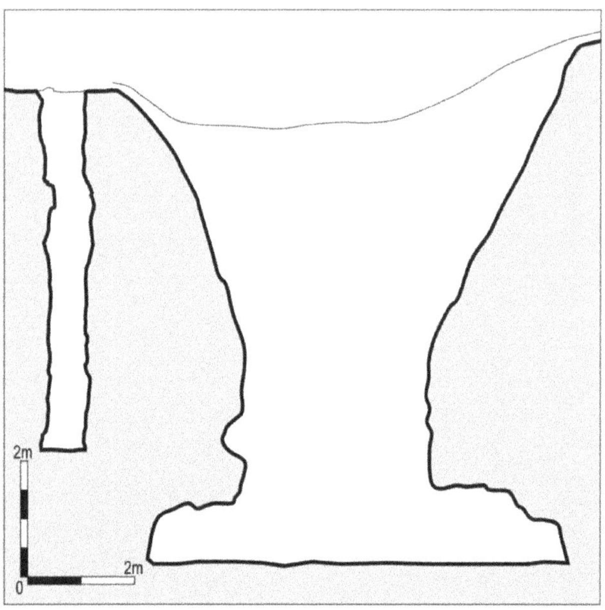

Fig. 10: A deep mining shaft from Casa Montero compared to scale with Grimes Graves Pit 1.

....

1 Compare, for example, with the 232 cm mean diameter of the excavated shafts at Jablines (Bostyn and Lanchon 1992).

The second observation refers to the main purpose of flint exploitations throughout Neolithic Europe. While many European mines are associated with the manufacture of flint axes, some with outstanding aesthetic qualities (e.g. Whittle 1995), the flint of Casa Montero – lacking all evident aesthetic quality– was dedicated to the production of blades. In fact, there is no evidence in Iberia of the use of flint for the manufacture of polished axes, mainly because of the wide variety of available hard stones such as sillimanite, ofite or amphibolite (e.g. Lillios 1997; Orozco 2000).

Prehistoric flint mines such as Casa Montero are particularly suitable archaeological sites for the analysis of how labour was organized and deployed by the first Neolithic societies. It is difficult to think of a site that materializes labour in a more direct and manifest form. There is absolutely no evidence of human activity other than flint mining. It was all production, with practically no consumption. Nevertheless, there is an aspect that should not be underestimated: the construction of monumental spaces and their use in the process of social reproduction. Unquestionably, one of the most relevant features of the Neolithic is a 'massive increase in the quantity of durable materiality' (Hodder 2005: 131). The mining process produced both use and probably exchange value, but also involved the progressive creation of a collective space. The first regional monumental space was erected as a result of the cumulative materialization of social labour. Generations of Neolithic miners created a durable monumental landscape while reinforcing the mechanisms of social reproduction involved in the transmission of communal rights over resources.

Marta Capote
Instituto de Historia
Centro de Ciencias Humanas y Sociales, CSIC
C/ Albasanz, 26-28, 28037, Madrid
e-mail: mcapote@ih.csic.es

Nuria Castañeda
Instituto de Historia
Centro de Ciencias Humanas y Sociales, CSIC
C/ Albasanz, 26-28, 28037, Madrid
e-mail: ncastanyeda@ih.csic.es

Susana Consuegra
Instituto de Historia
Centro de Ciencias Humanas y Sociales, CSIC
C/ Albasanz, 26-28, 28037, Madrid
e-mail: sconsuegra@ih.csic.es

Cristina Criado
Instituto de Historia
Centro de Ciencias Humanas y Sociales, CSIC
C/ Albasanz, 26-28, 28037, Madrid
e-mail: cctorija@ih.csic.es

Pedro Díaz-del-Río
Instituto de Historia
Centro de Ciencias Humanas y Sociales, CSIC
C/ Albasanz, 26-28, 28037, Madrid
e-mail: diazdelrio@ih.csic.es

REFERENCES

Barber, M., Field, D. and Topping, P. 1999. The Neolithic Flint Mines of England. Swindon: English Heritage.

Bosch, J. and Strada, E. 1994. La Venus de Gavá (Barcelona). Una aportación fundamental para el estudio de la religión neolítica en el suroeste europeo. Trabajos de Prehistoria 51(2), 149-158.

Bostyn, F. and Lanchon, Y. (eds.) 1992. Jablines. Le Haut Château (Seine-et-Marne). Une minière de silex au Néolithique. Documents d'Archéologie Française, 35. Paris: Editions de la Maison des Sciences de L'Homme.

Bostyn, F. and Lanchon, Y. 1997. The Neolithic Flint Mine at Jablines, 'Le Haut Château' (Seine-et-Marne). In A. Ramos-Millán and M.A. Bustillo (eds.), Siliceous Rocks and Culture, Granada : Universidad de Granada,271-291.

Brell, J. M., Doval, M. and Caramés, M. 1985. Clay mineral distribution in the Evaporitic Miocene Sediments of the Tajo Basin, Spain. Minealogica et Petrographica Acta 29-A, 267-276.

Burton, J. 1984. Quarrying in a Tribal Society. World Archaeology16 (2), 234-247.

Bustillo, M. A., Castañeda, N., Capote, M., Consuegra, S., Criado, C., Díaz-del-Río, P., Orozco, T., Pérez-Jiménez, J. L. and Terradas, X. 2008: Is the macroscopic classification of flint useful ? A petroarchaeological analysis and characterization of flint raw materials from the Iberian Neolithic mine of Casa Montero. Archaeometry, DOI: 10.1111/j.1475-4754.2008.00403.x.

Bustillo, M. A. and Pérez-Jiménez, J. L. 2005. Características diferenciales y génesis de los niveles silíceos explotados en el yacimiento arqueológico de Casa Montero (Vicálvaro, Madrid). Geogaceta 38, 243-246.

Bueno Ramírez, P., Barroso Bermejo, R., Balbín Behrman, R., Campo Martín, M., Etxeberría Gabilondo, F., González Martín, A., Herrasti Erlogorri, L., Treserras, J. J., López García, P., López Sáez, J. A., Matamala, J. C. and Sánchez, B. 2002. Áreas habitacionales y funerarias en el Neolítico de la cuenca interior del Tajo: La provincia de Toledo. Trabajos de Prehistoria 59 (2), 65-79.

Camprubí, A., Melgarejo, J-C., Proenza, J.A., Costa, F., Bosch, J., Estrada, A., Borell, F., Yushkin, N.P. and Andreichev, V.L. 2003. Mining and geological knowledge during the Neolithic: a geological study on the variscite mines at Gavà, Catalonia. Episodes 26(4), 295-301.

Capote, M., Castañeda, N., Consuegra, S., Criado, C., Díaz-del-Río, P., Bustillo, M. A. and Pérez-Jiménez, J. L. 2006. Casa Montero, la mina de sílex más antigua de la Península Ibérica. Tierra y Tecnología 29, 42-50.

Castañeda, N., Capote, M., Criado, C., Consuegra, S., Díaz-del-Río, P., Terradas, X. and Orozco, T. 2008. Las cadenas operativas líticas de la mina de sílex de Casa Montero (Madrid). IV Congreso del Neolítico en la Península Ibérica (Alicante, 27-29 de Noviembre de 2006). Volume 2, Alicante: Museo Arqueológico de Alicante, 231-234.

Castañeda, N. and Criado, C. 2006. La industria lítica de Casa Montero (Vicálvaro, Madrid): resultados preliminares. In N. Ferreira and H. Verissimo (eds.). Do Epipaleolítico ao Calcolítico na Península Ibérica. Actas do IV Congresso de Arqueologia Peninsular (Faro, 14 a 19 Setembro de 2004), Faro, 229-234.

Consuegra, S., Castañeda, N. and Gallego, M. M. 2005. Explotación minera de sílex neolítica de Casa Montero (Vicálvaro, Madrid). Avance de los resultados de las excavaciones. In O. Puche and M. Ayarzagüena (eds.), Minería y metalurgia histórica en el sudoeste Europe, 45-51. Madrid.

Consuegra, S. Gallego, M. M. and Castañeda, N. 2004. Minería neolítica de sílex de Casa Montero (Vicálvaro, Madrid). Trabajos de Prehistoria 61(2), 127-140.

Delibes, G., Rojo, M. and Represa, I. 1993. Dólmenes de La Lora: Burgos. Junta de Castilla y León. Consejería de Cultura y Turismo.

Díaz-del-Río, P. 2004. Copper Age ditched enclosures in Central Iberia. Oxford Journal of Archaeology 23 (2), 107-121.

Díaz-del-Río, P. and Consuegra, S. 1999. Primeras evidencias de estructuras de habitación y almacenaje neolíticas en el entorno de la Campiña madrileña: el yacimiento de 'La Deseada (Rivas-Vaciamadrid, Madrid). In J. Bernabeu Aubán and T. Orozco Köhler (eds) II Congrés del Neolític a la Península Ibèrica, 251-257. València: Saguntum Extra.

Díaz-del-Río, P., Consuegra, S., Capote, M., Castañeda, N., Criado, C., Orozco, T. and Terradas, X. 2008. Estructura, contexto y cronología de la mina de sílex de Casa Montero (Madrid). IV Congreso del Neolítico en la Península Ibérica (Alicante, 27-29 de noviembre de 2006). Volume 1, Alicante: Museo Arqueológico de Alicante, 200-207.

Díaz-del-Río, P., Consuegra, S., Castañeda, N., Capote, M., Criado, C., Bustillo, M. A. and Pérez-Jiménez, J. L. 2006. The earliest flint mine in Iberia. Antiquity 80 (307). http://antiquity.ac.uk/ProjGall/diazdelrio/index.html

Di Lernia, S., Fiorentino, G., Galiberti, A. and Basili, R. 1995. The Early Neolithic mine of Defensola 'A' (I 18): flint exploitation in the Gargano area. Archaeologia Polona 33, 119-132.

Estremera, M. S. 2003. Primeros agricultores y ganaderos en la Meseta Norte: el Neolítico de la Cueva de la Vaquera (Torreiglesias, Segovia). Arqueología en Castilla y León. Consejería de Educación y Cultura. Junta de Castilla y León.

Felder, P. J. S. 1997. Excavations of Prehistoric Flint Mines at Ryckholt-St. Geertruid (Netherlands), Grimes Graves and Harrow Hill (United Kingdom). In A. Ramos-Millán and M.A. Bustillo (eds), Siliceous Rocks and Culture, 143-150. Granada: Universidad de Granada.

Galiberti, A., Di Lernia, S., Florentino, G. and Guariscio, M. 1997. New Data on The Neolithic Flint Mine of Defensola, Vieste (Italy). In A. Ramos-Millán and M.A. Bustillo (eds.), Siliceous Rocks and Culture, 211-222. Granada: Universidad de Granada.

Galiberti, A., Sivilli, S. and Tarantini, M. 2001. La miniera neolitica della Defensola (Vieste-Foggia): Lo statu delle ricerche. Origini XXIII, 85-110.

Hodder, I. 2005. The spatio-temporal organization of the early 'town' at Çatalhöyük. In D. Bailey, A. Whittle and V. Cummings (eds), (Un)settling the Neolithic, 126-139. Oxford: Oxbow.

Holgate, R. 1995. Neolithic flint mining in Britain. Archaeologia Polona 33, 133-161.

Junco, F. and Calvo, J. P. 1983. Cuenca de Madrid. In W. Gibbons and T. Moreno (eds), Geología de España. IGME, Tomo II, 534-543.

Kunst, M. and Rojo, M. 1999. El Valle de Ambrona: un ejemplo de la primera colonización Neolítica de las tierras del Interior Peninsular. In J. Bernabeu Aubán and T. Orozco Köhler (eds), II Congrés del Neolític a la Península Ibèrica, 259-270. València: Saguntum Extra.

Lech, H. and Lech, J. 1984. The Prehistoric Flint Mine at Wierzbica 'Zele': A Case Study from Poland. World Archaeology 16(2), 186-203.

Lech, H. and Lech, J. 1997. Flint Mining among Bronze Age Communities: A Case Study from Central Poland. In R. Schild and Z. Sulgostowska (eds), Man and Flint. Proceedings of the VIIth International Flint Symposium. Warszawa-Ostrowiec Świętokrzyski, 91-98. Warszawa: Institute of Archaeology and Ethnology Polish Academy of Sciences.

Lillios, K.T. 1997. Amphibolite tools of the Portuguese Copper Age (3000-2000 BC): a geoarchaeological study of prehistoric economics and symbolism. Geoarchaeology 12(2), 137-163.

Mercader, J., Cortés, A.F. and García, M.E. 1989. Materiales neolíticos en el valle del Jarama (Arganda, Madrid). Trabajos de Prehistoria 46, 255-260.

Migal, W. and Salacinski, S. 1997. Studies at Krzemionki during the last decade. In R. Schild and Z. Sulgostowska, Man and Flint. Proceedings of the VIIth International Flint Symposium. Warszawa-Ostrowiec Świętokrzyski, 103-108. Warszawa: Institute of Archaeology and Ethnology Polish Academy of Sciences.

Orozco, T. 2000. Aprovisionamiento e Intercambio: Análisis petrológico del utillaje pulimentado en la Prehistoria Reciente del Pais Valenciano (España). B. A. R. Int. Ser. S867, Oxford: BAR.

Pascual-Benito, J. L. 1996. Los anillos neolíticos de la Península Ibéria, Congrés del Néolitic a la Península Ibèrica (Gavá-Bellaterra). Rubricatum I, 279-289.

Pérez-Jiménez, J.L., Alonso-Zarza, A.M. and Bustillo, M.A. 2005. Sedimentología y diagénesis de las secuencias lacustres someras en el área del yacimiento neolítico de Casa Montero (Vicálvaro, Madrid). Geo-Temas 8, 77-80.

Prado, C. de, 1864. Descripción física y geológica de la provincia de Madrid. Madrid: Imprenta Nacional.

Ramos-Millán, A. 1997. La Venta. A prehistoric Flint Mine in a Tribal Society (Iberian South East).I n R. Schild and Z. Sulgostowska (eds), Man and Flint. Proceedings of the VIIth International Flint Symposium. Warszawa-Ostrowiec Świętokrzyski, Warszawa: Institute of Archaeology and Ethnology Polish Academy of Sciences, 117-121.

Ramos-Millán, A., Osuna-Vargas, M.M., Tapia-Espinosa, A., Pena-González, B. and Aznar-Pérez, J.C. 1997. Archaeological Research in the La Venta Flint Mine (Iberian Southeast). In A. Ramos-Millán and M.A. Bustillo (eds), Siliceous Rocks and Culture, Granada: Universidad de Granada, 245-270.

Rodríguez González, D. 2006. Historia de la investigación y estado de la cuestión del Neolítico en Castilla-La Mancha: una visión de conjunto. In N. Ferreira y H. Verissimo (eds), Do Epipaleolítico ao Calcolítico na Península Ibérica. Actas do IV Congresso de Arqueologia Peninsular (Faro, 14 a 19 Setembro de 2004), 217-228.

Rojo, M. and Kunst, M. 1999. La Lámpara y la Peña de La Abuela. Propuesta secuencial del Neolítico Interior en el ámbito funerario. In J. Bernabeu Aubán and T. Orozco Köhler (eds), II Congrés del Neolític a la Península Ibèrica, València: Saguntum Extra, 503-512.

Rojo, M., Kunst, M., Garrido, R., García, I. and Morán, G. 2005. Un desafío a la eternidad: tumbas monumentales del valle del Ambrona, Soria. Memorias de Arqueología en Castilla y León, 14, Consejería de Cultura y Turismo. Valladolid : Junta de Castilla y León.

Schild, R. 1995. Tomaszów, Radom Province. Archaeologia Polona, 33, 455-465.

Stout, D. 2002. Skill and Cognition in Stone Tool Production. An Ethnographic Case Study from Irian Jaya. Current Anthropology 43(5), 693-722.

Topping, P. 1997. Structured deposition, symbolism, and the English flint mines. In R. Schild and Z. Sulgostowska (eds), Man and Flint. Proceedings of the VIIth International Flint Symposium. Warszawa-Ostrowiec Świętokrzyski, Warszawa: Institute of Archaeology and Ethnology Polish Academy of Sciences,127-131.

Yravedra, J., Maicas, R., Díaz-del-Río, P. and Consuegra, S. 2008. Anillos para una mina. Industria ósea y fauna de la mina de sílex neolítica de Casa Montero (Madrid). IV Congreso del Neolítico en la Península Ibérica (Alicante, 27-29 de noviembre de 2006). Volume 2, Alicante: Museo Arqueológico de Alicante, 240-247.

Whittle, A. 1995. Gifts from the earth: symbolic dimensions of the use and production of Neolithic flint and stone axes. Archaeologia Polona 33, 247-259.

INTENSIVE EXTRACTION OF NON-METALLIC MINERALS DURING THE EARLY PROTOHISTORY IN THE NORTHERN HALF OF EUROPE

Yoann Gauvry

Abstract: The main aim of this work is, firstly, to determine the state of knowledges about mineral exploitation available in archaeological literature, in order to study, secondly, problematics elaborated on a geographically and chronologically large scale. The techniques for acquiring non-metallic mineral raw materials, which include every degrees between ponctual surface gathering untill deep and organized mines, are full of informations about the people who employed them ; it is interesting to understand their evolution during the different periods of Neolithic, Chalcothic and early Metal Ages, and to see if they obey global and common rules in the northern half of Europe.

Keywords: Mines, quarries, stone, flint, raw material extraction, Neolithic, Chalcolithic, Europe.

INTRODUCTION

This paper will present the main issues and results of a general study on the contribution of mining archaeology to our understanding of later Prehistory, especially during the Neolithic and Chalcolithic periods. It concerns the northern half of Europe, and is conceived in the light of previous well known studies on a global scale, such as those of Clark and Pigott 1933, Jahn 1960, Schmid 1973 and 1974, Lech 1981a-b... Each of these studies has shown the importance of replacing the data on stone exploitation within a broader framework, both geographical and chronological. This is essential in order to grasp their full historical and cultural potential.

The geographical scope of this review covers the northern half of Europe, from the dawn of the Neolithic until the beginning of the Bronze Age (Fig. 1). All the published data regarding the archaeology of the principal stone mines and quarries have been collected; a corpus of a hundred and ten exploitation sites has been brought together, which constitutes the basis of this overall study (Fig. 2).

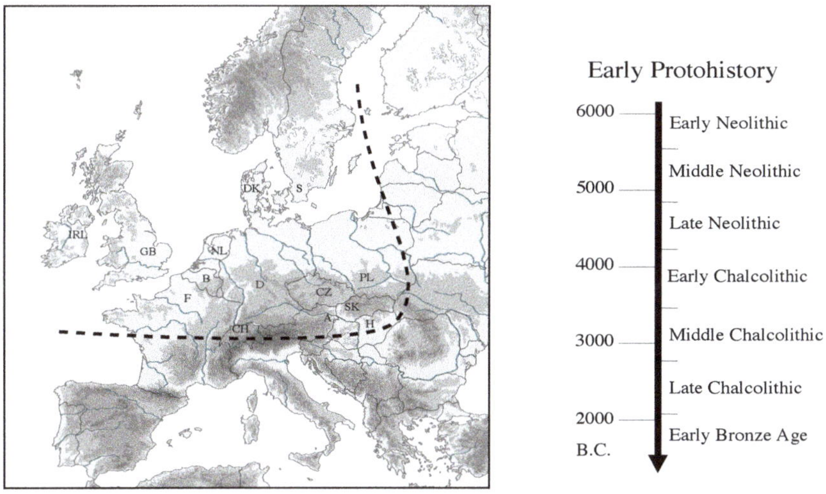

Fig. 1: Geographical and chronological frame of the study

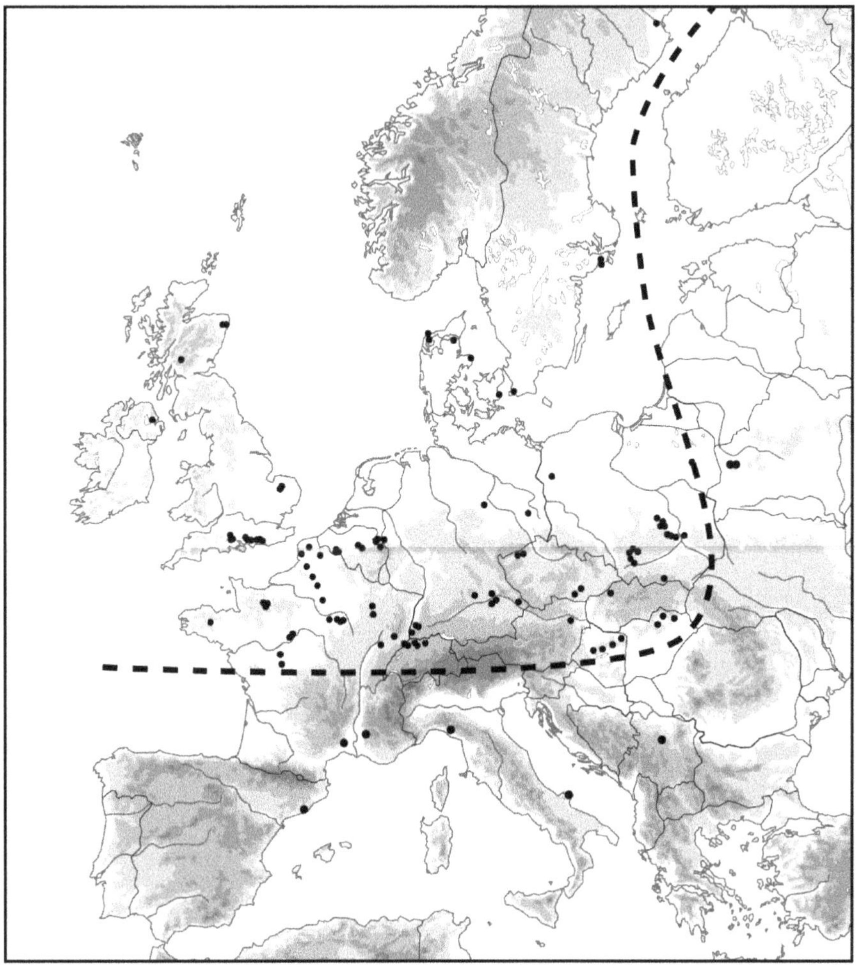

Fig. 2: Localisation of the 110 sites treated in the study

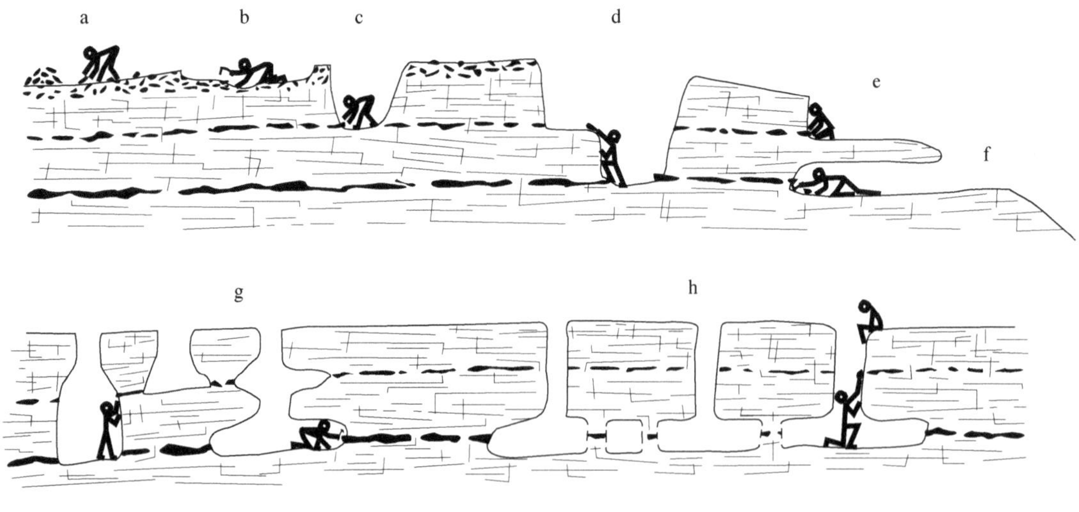

Opencast exploitation
a : surface gathering
b : surface quarrying
c : shallow quarrying
d : deep quarrying

Slope exploitation
e : slope quarrying
f : slope tunneling

Underground exploitation - mining
g : underground extension
h : underground network

Fig. 3: the main exploitation technics, illustrated by flint extraction.

Among the several hundred references identified during the past 160 years, these particular sites have been selected on the basis of the good, or at least acceptable, quality of information concerning them. The corpus includes sites of all sizes, depths, densities; they have been the object of many excavations, which has made it possible to recognize various types of exploitation techniques: opencast mining and underground galleries, simple and complex sites, rudimentary or well organized production techniques (Fig. 3)…

Studying stone mines and quarries requires that one takes into account several specific difficulties, which occur to all levels of interpretation. The complexity of the formation of exploitation structures and of their fill, as well as their frequent lack of artefactual material, makes understanding and dating them difficult. Moreover, it is rarely possible to establish clear associations between particular extraction techniques and precisely dated cultural complexes. Taking these difficulties into account, we can nonetheless propose three principal stages in such a study:

-First: to compare these sites with one another, and to construct a typology of the extraction methods and techniques ;

-Second: to deduce how technical procedures were developed in response to socio-economical demands;

-Third: to place these data along a chronological axis, and translate them into chrono-cultural interpretations.

Thus, the ultimate objective of this work can be summarized by these three questions:

-When and how did intensive mining of stone appear in Europe?

-How did it develop?

-What does it teach us about early protohistory periods?

1. CHRONOLOGICAL REVIEW

1.1. Definition

In order to treat a global question of early protohistory, it is necessary to adopt large contemporary horizons to divide this era. Here, the early protohistory subdivision and terminology defined in the *Protohistoire de l'Europe* (Lichardus *et al.* 1985) appears to us to be the most appropriate choice. Therefore, unless it is specifically mentioned, all the chronological stages of early protohistory (late Neolithic, middle Chalcolithic, etc.) must be understood in this pan European meaning.

1.2. Neolithic

There is little positive evidence of Neolithic exploitation structures (Fig. 4). In many mines and quarries, some Neolithic material and absolute dates have been identified; but they are rarely clearly related to exploitation structures.

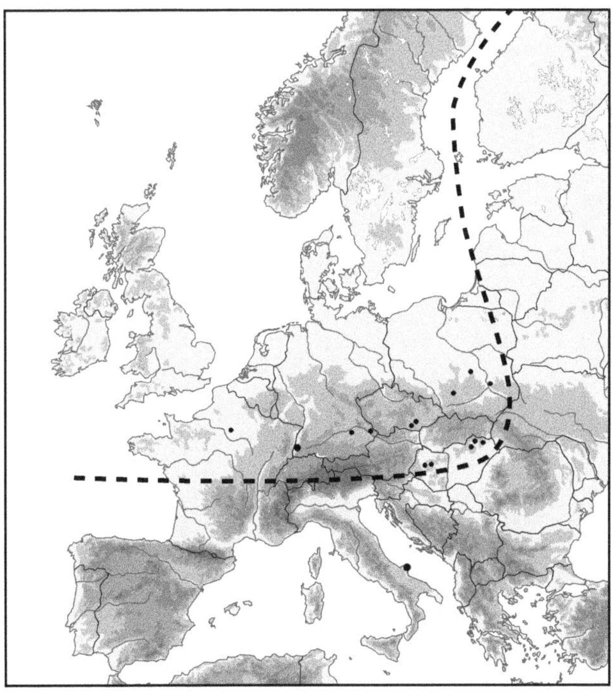

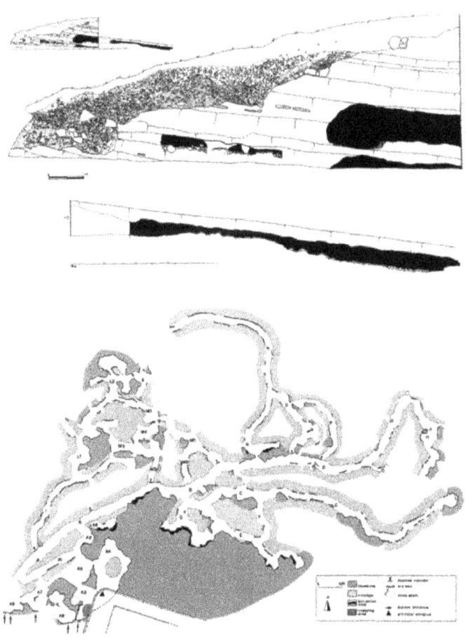

Exploitaition structures of La Defensola (Italy)
(Di Lierna *et al.* 1995: 121-122)

Fig. 4 : exploitation sites known and supposed for the middle and late Neolithic.

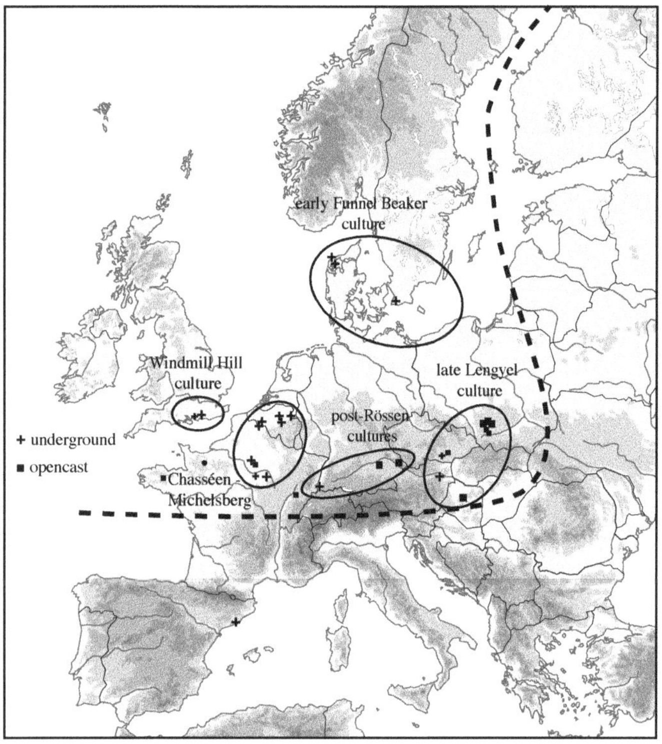

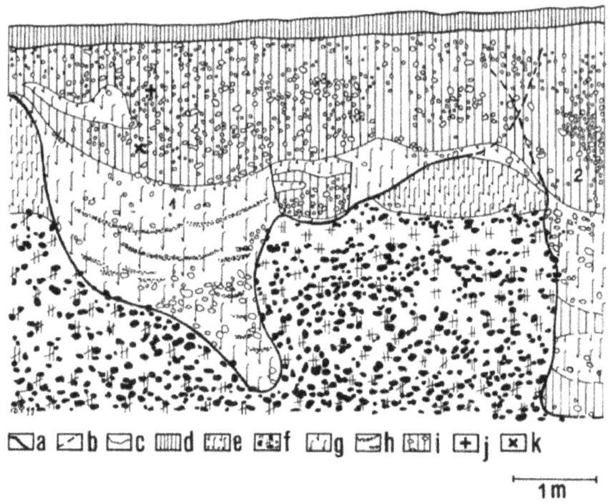

Deep quarrying in Saspow (Poland) (Lech in Babel *et al.* 1999: 618)

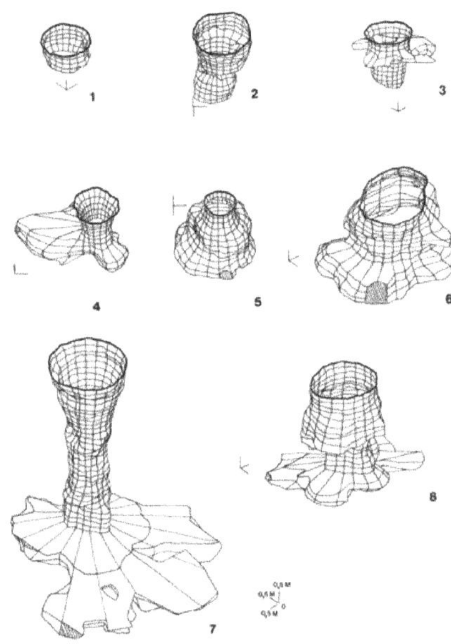

Opencast and underground structures in Jablines (France) (Bostyn and Lanchon 1992: 213)

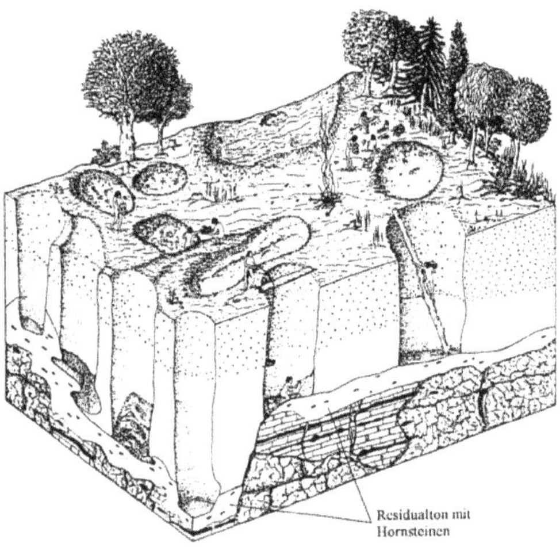

Deep quarrying in Arnhofen (Germany) (Gayck 2000: 270)

Underground extension in Hov (Denmark) (Becker and Weisgerber 1999: 463)

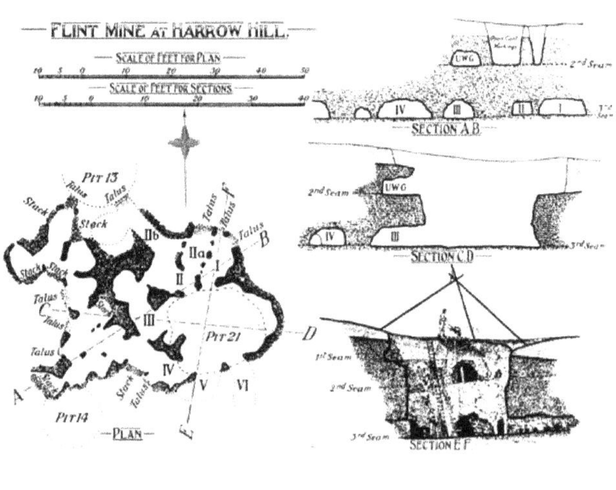

Underground network in Harrow Hill (Great-Britain) (Shepherd *et al.* 1999: 521)

Fig. 5 : exploitation sites of early Chalcolithic, and details.

When material has been found on the surface of a site, it is not necessarily possible to establish a direct relation with the exploitation structures. The artefacts might be related to a previous exploitation phase, in which entirely different techniques were employed. Moreover, most of the raw materials employed during the Neolithic epoch were also extracted during the following era (Chalcolithic), and the primary structures masked or destroyed by the later extraction phases. For example, we know that Jurassic Krakow flint was habitually used by Neolithic societies, but the main exploitation sites must be attributed to Chalcolithic times, according to several radiocarbon datings (Lech 1981a, 48) and to cultural characteristics as well (Dzieduszycka-Machnikowa and Lech 1976: 167-168; Lech 1981b: 275).

By way of illustration, here are a few interesting cases. The ochre exploitation structures of Bad Sulzburg (Black Forest, Germany), excavated in the late 1980s, provided several radiocarbon calibrated datings around 5000 BC (Goldenberg *et al.* 2003: 185). However, the similarity between these exploitation structures and tools and those of the jasper exploitation at Kleinkems, very close to Bad Sulzburg, and dated 1000 years later, invites caution about attributing this site to the Neolithic.

In the reliably dated Italian mine of La Defensola, flint was extracted in large galleries during early and middle Neolithic. This is a particularly interesting site. The exploitation was carried out through galleries 60 cm high on average, which formed a network of more than 6000 m^2 on the level of the higher bed. The waste products were contained behind low dry stone walls, carefully arranged to preserve a clean and safe passage. Ceramics and many extraction tools were found on the exploitation floor; this material, in addition to several radiocarbon datings, indicates a Neolithic context (Galiberti *et al.* 1997: 220).

La Defensola, although outside of our geographical research area, raises very exciting questions about the oldest indisputably underground mines known in Europe. The same questions can also be asked about the Spanish flint mine of Casa Montero, dated around 5400-5200 cal BC, which shows all the characteristics of the later mines (see capote *et al.*, this volume). These two exceptions highlight the need to extend our study to the southern half of Europe, and eventually to open it to the Mediterranean context.

1.3. Early Chalcolithic

A great many of the best known and most spectacular mines and quarries are attributed to the early Chalcolithic (Fig. 5). However it is important to recall that throughout the Chalcolithic, these great mines and quarries coexisted with small and short-lived exploitations, which are attested at every phase of the periods under review. We will conduct this review from East to West, following the major cultural groups.

Late Lengyel culture

Mineral resources attract populations, who settle in their vicinity in spite of soils most unfavourable to agriculture. This fact is evident at the site of Szentgál (Transdanubia, Hungary), where mineral resources attracted populations who settled nearby, despite poor agricultural terrain (Biró and Regenye 2003: 59-60).

This phenomenon also occurs in regard to the Jurassic Krakow flint, which played a major role ahead during Neolithic times and during the Chalcolithic, when important deposits of this material attracted new groups of people to previously uninhabited areas (Lech 1981b: 275). Important exploitation sites developed, such as at Saspów (wojw. Kraków, Poland), where an intensive extraction by deep quarrying was set up to obtain nodules in sub-original positions in a layer of clay several meters deep. Raw material was not only obtained, but was also partially transformed on the site. The production steps with the highest risk of failure, and creating the largest quantity of waste, was carried out on the site in order to limit the amount of non-useful material to be transported (Dzieduszycka-Machnikowa and Lech 1976).

On the Vienne-Mauer (Austria) site, the extraction of radiolarite in much degraded chalk was carried out in shafts which reached a depth of 6 to 10 m, interconnected by horizontal galleries (Ruttkay 1970). A feature of particular interest here is a burial site with seven individuals, men, women and children, probably linked to the mining activities.

Post-Rössen cultures

A number of 'Jurahornstein' deposits are found throughout Bavaria, on both sides of the Danube. This raw material was used from the middle Neolithic on, but it is difficult to specify the exploitation techniques employed at that time. We can mention the large shallow pits of Flintsbach-Hardt, and the deep shafts of Arnhofen (Bavaria, Germany) in which silexite nodules were extracted from a thick layer of surface clay. Unfortunately these structures are imprecisely dated, as the raw materials were probably extracted at different periods in the same place. Several elements seem to indicate that the principal phases of mining took place during the post-Rössen and Münchshöfen cultures (Eisele *et al.* 2003: 61).

On the Kleinkems (Baden-Württenberg, Germany) site, the slope quarrying and slope mining structures and the tools were very similar to those found in the ochre exploitation in the Black Forest, which is, however, 1000 years older according to radiocarbon dating. Fire was used to fragment the chalk, which was then ground up with heavy quartzite hammers in order to free up the beds of jasper. Unfortunately, we don't know how the material was used in the region. This exploitation was fairly intensive. The site was clearly dated by radiocarbon samples and by ceramic shards present in two tombs clearly attributed to the Munzingen group (Schmid 1952: 12; Engel and Sigmund 2005).

Michelsberg / Chasséen

The Michelsberg and Chasséen cultures are well documented, and include some of the most spectacular mines.

The Rijckholt-St. Geertruid (Limburg, Netherlands) site, with its extensive underground networks, illustrates extremely rational and efficient exploitation techniques for a deposit whose depth varies between 2 and 12 m (Felder *et al.* 1998). The transformation of the raw material, essentially during the stages carrying the highest risk of failure was partly carried out on site: roughouts of axes and and pre-core shaping, debitage of blades, before transport away from the site to finish off the production. These products played an important regional role during the Michelsberg culture.

In the present state of knowledge, the record depth level has been observed at Spiennes (Hainaut, Belgium), with shafts 16 m deep at Camp-à-Cayaux. There is a certain contrast between the different exploitation areas, which for the most part were in activity in the Michelsberg culture (Collet 2004: 131).

As for the Chasséen, many important sites are known. The Jablines site (Seine-et-Marne, France) shows a great variety of structures, from shallow pits to shafts 7 m deep with an extremely rational organisation so as to best cover the deposit (Bostyn and Lanchon 1992). We should also mention the Villemaur-sur-Vanne 'Les Orlets' site (Aube, France), where four flint beds were exploited in structures very close together, sometimes overlaying one another: a clear illustration of the geographical concentration of vestiges from different exploitation phases (De Labriffe *et al.* 1995: 110).

North-western Europe cultures

We note that the appearance of intensive extraction concerns the whole continent, even the peripheral regions of north-western Europe, which had not yet been reached by the Neolithic way of life.

For example, underground mining is a feature of the early Funnel Beaker culture (FBC) in southern Scandinavia, as can be seen at Kvarnby (Malmö district, Suede), or at Hov (Thisted district, Denmark). The intensive exploitation of flint beds here can reach down to 8 m, with wooden structures detectable in the shaft fills. In England, the underground flint exploitation was used from the start, cf. the Windmill Hill culture. This is clearly visible on the South Downs sites, particularly at Harrow Hill and Long Down (West Sussex, Great Britain).

1.4. Middle Chalcolithic

Thereafter, the main tendencies observed in early Chalcolithic seem to continue during the middle Chalcolithic, even though we have less information concerning that period (Fig. 6).

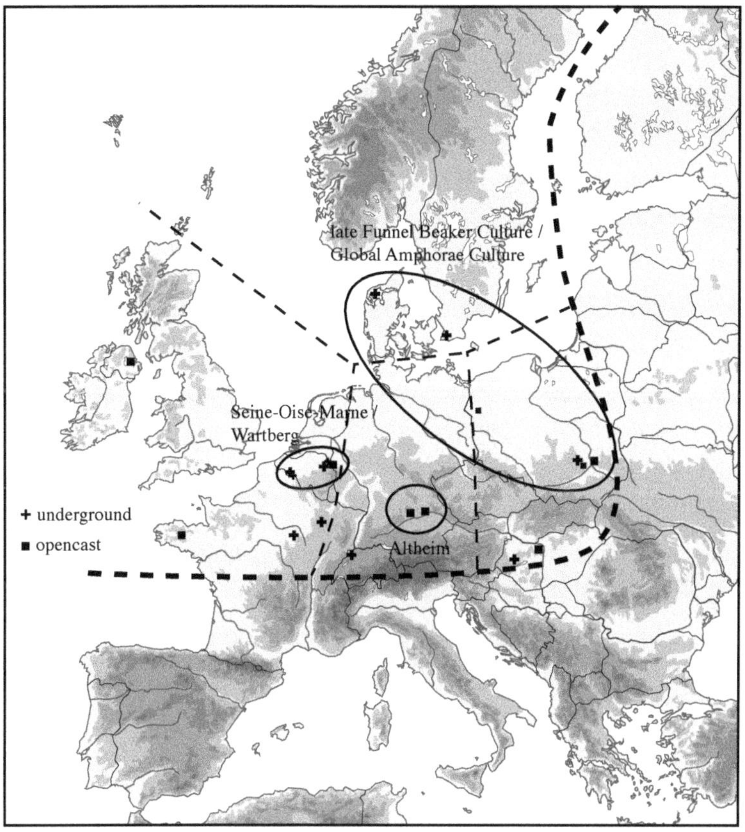

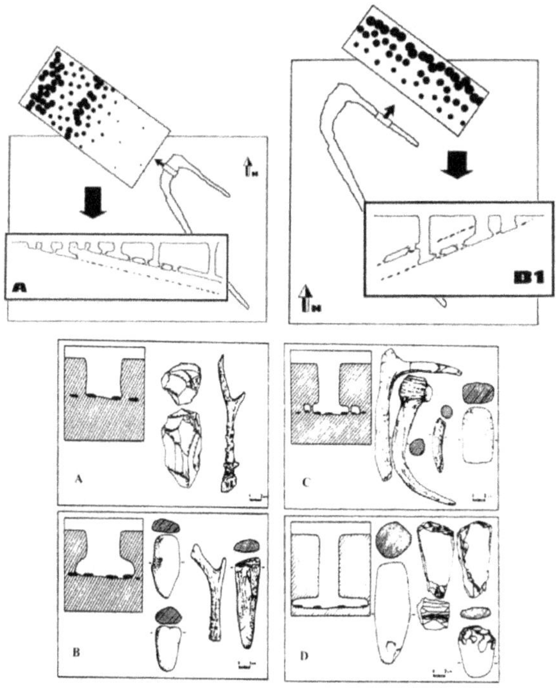

Exploitation structures and tools in Krzemionki (Poland) (Migal 1997: 101 ; Borkowski 1997: 333)

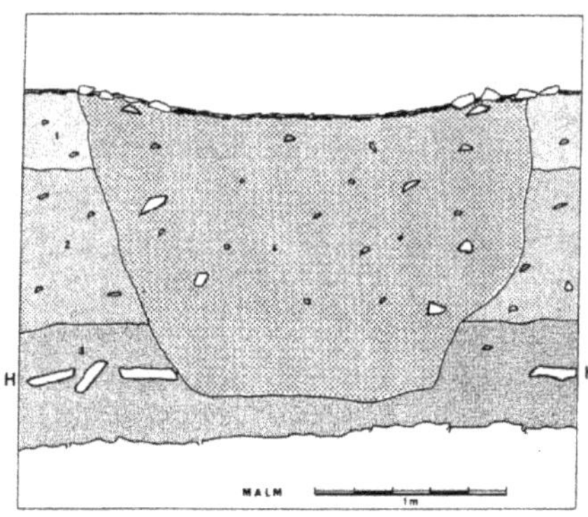

Shallow quarrying in Scernfelder Forst (Germany) (Gayck 2000: 284)

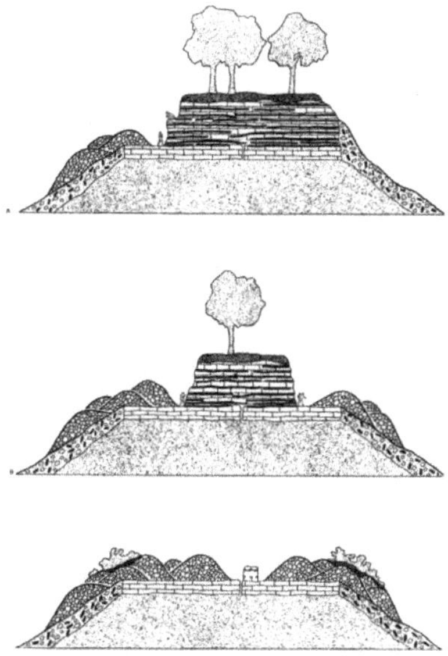

Great slope quarrying in Lousberg (Germany) (Weiner and Weisgerber 1999: 105)

Fig. 6 : exploitation sites of the middle Chalcolithic, and details.

Late FBC

In Poland, we observe a few changes compared to the preceding period: after the predominance of the Jurassic Krakow flint and chocolate flint, we note the important role played by the flint extracted at Swieciechow (wojw. Tarnobrzeg, Poland). This high quality raw material is found at less than a meter below the surface, and spreads out over a large area (570 km) in the form of axes and blades (Balcer 1976: 190). It hides the appearance of the Krzemionki striped flint site at the beginning of the middle Chalcolithic, first exploited in opencast structures.

It is important to note that at the Gorzsów site (wojw. Gorzsów Wielkopolski, Poland), flint nodules were gathered in the gravel of a small river, and used directly in a local settlement (Bąbel *et al.* 1999: 624). This type of exploitation is very light, leaving few traces behind, and must have been quite common over the whole of early protohistory, satisfying small-scale demands other than those that led to developing mines of more consequence.

At the end of middle Chalcolithic, during the Globular Amphorae culture (GAC), we observe the acme of striped flint exploitation at Krzemionki (wojw. Tarnobrzeg, Poland) (Migal 1997b: 324). There, the variety of the deposit layout entailed a large spectrum of extraction techniques, in order to obtain a raw material of high quality, favoured for its attributes both mechanical and aesthetic.

Altheim culture

Widespread use of 'Jurahornstein' continued in Bavaria, but the exploitation techniques of this period are poorly documented and dated. There are indications that raw material was scarce and intensively extracted at sites such as Baiersdorf and Schernfleder Forst (Bavaria, Germany).

Northwestern Europe

During the middle Chalcolithic, particularly in the Wartberg culture, Lixhe-Lanaye flint extracted at Rijckholt-St. Geertruid is still commonly found, obtained in the same structures as during the previous period. However, other raw materials played a more important role, for example Valkenburg flint, extracted nearby. Another source of growing importance was the flint mined in the Lousberg hill (Ldkr. Aachen, Germany), through a technique of large-scale slope levelling which almost razed the 4 upper meters of the hill. Finally, we note the continuation of mining activities at Spiennes during the Seine-Oise-Marne culture (SOM), principally in the Petit-Spiennes aera, through underground galleries not so deep as earlier (Collet *et al.* 2004: 157).

1.5. Late Chalcolithic

The late Chalcolithic era is not so well documented (Fig. 7). However, several global tendencies may be observed, specifically in the peripheral regions of Europe.

In most of the sites, modest, opencast and shallow pit-mining was common, with the raw material locally disseminated and used. Some of these had been intensively mined during earlier periods, chocolate flint for instance, but during the late Chalcolithic these products are less frequently found.

We do however know of some intensive exploitation sites. Among these the sites of Krasnoye Selo and Karpautsy (Belarus), and especially Grimes Graves (Norfolk, Great Britain), where we find spectacular underground networks. These few examples of intensive underground mineral extraction are geographically near major mining sites of ancient and middle Chalcolithic.

Finally, some more modest exploitation sites display specific characteristics: they are linked to the production of new types of artefacts of non-functional character. Take, for example, the Danish and Swedish flint daggers (Becker 1959: 88), they were made from raw material obtained through shallow and simple structures such as the ones of Aalborg (Denmark), and even collected on coastal beaches such as Fornaes (Denmark). The production of Grand Pressigny flint artefacts in France obeyed similar rules: one notes the same contrast between simple and modest extraction techniques and the production of elegant 40 cm long blades requiring very sophisticated individual know-how (Pelegrin 2002), and which seem intended more for ceremonial than for practical use.

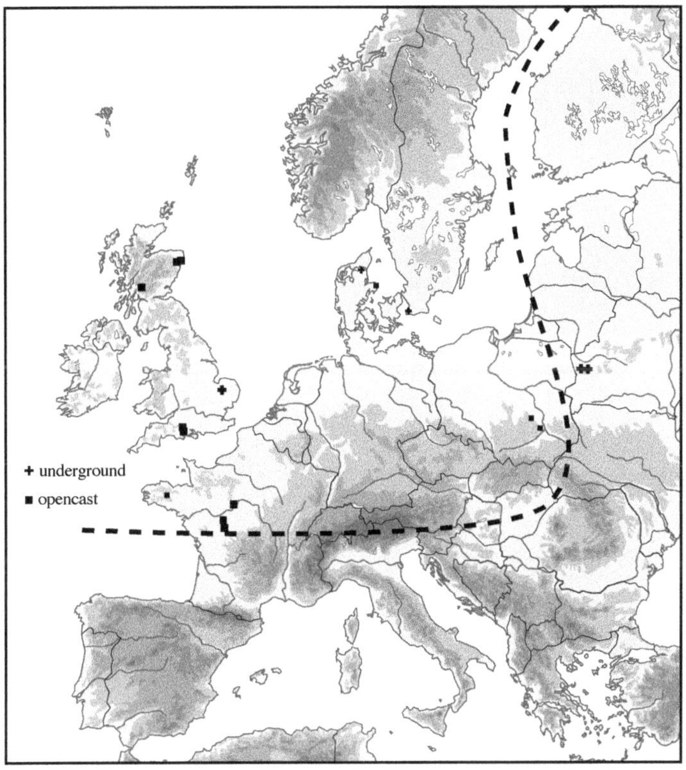

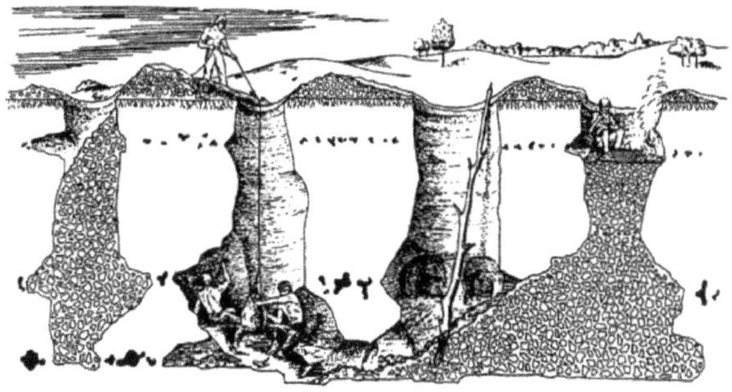

Underground network in Krasnoye Selo (Belarus) (Gurina 2000: 86)

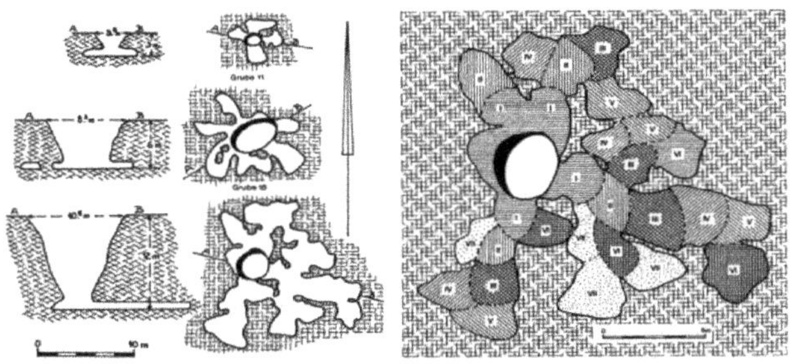

Underground structures in Grimes Graves (Great Britain) (Felder 1999: 121)

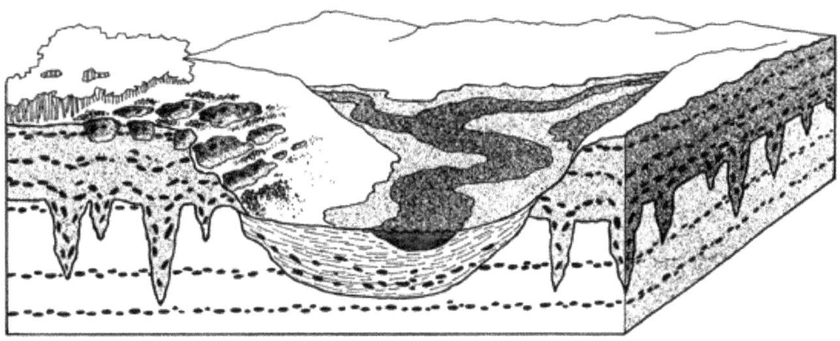

Opencast structures supposed for the extraction of Grand Pressigny flint (France) (Dijkman *et al.* 1999: 484)

Fig. 7 : exploitation sites of late Chalcolithic, and details.

1.6. Early Bronze Age and after

Stone extraction continued to decline during the early Bronze Age, when metal production came to play the major role (Fig. 8). However, in some areas distant from the main axis of metal diffusion, some kinds of stone extraction survived. For example, several opencast and underground extraction sites have been attributed to the culture of Mierzanowice, in Poland.

In several mining sites of Central and Western Europe, such as Rijckholt-St. Geertruid and Grimes Graves, some Iron Age occupation is attested both by ceramics and by absolute dating techniques. Such artefacts seem episodic and slight, compared to the extensive and regular extraction of the previous periods.

2. INTERPRETATION

2.1. The Neolithic-Chalcolithic transition

Through technological, petro-archaeological and functional analyses, we know that Neolithic stone economies were in various ways complex and geographically extensive, involving a large spectrum of strategies to obtain raw material. Except in a couple of cases, exploitation techniques used then are not well known; we may suppose that pit-mines were the rule, and that these extraction activities were regularly kept up, but not with the same intensity as during the later periods. We may assume that a Neolithic 'substratum' largely determined subsequent development. It already included most of the features which characterised, to various degrees, the 'stone economics' of

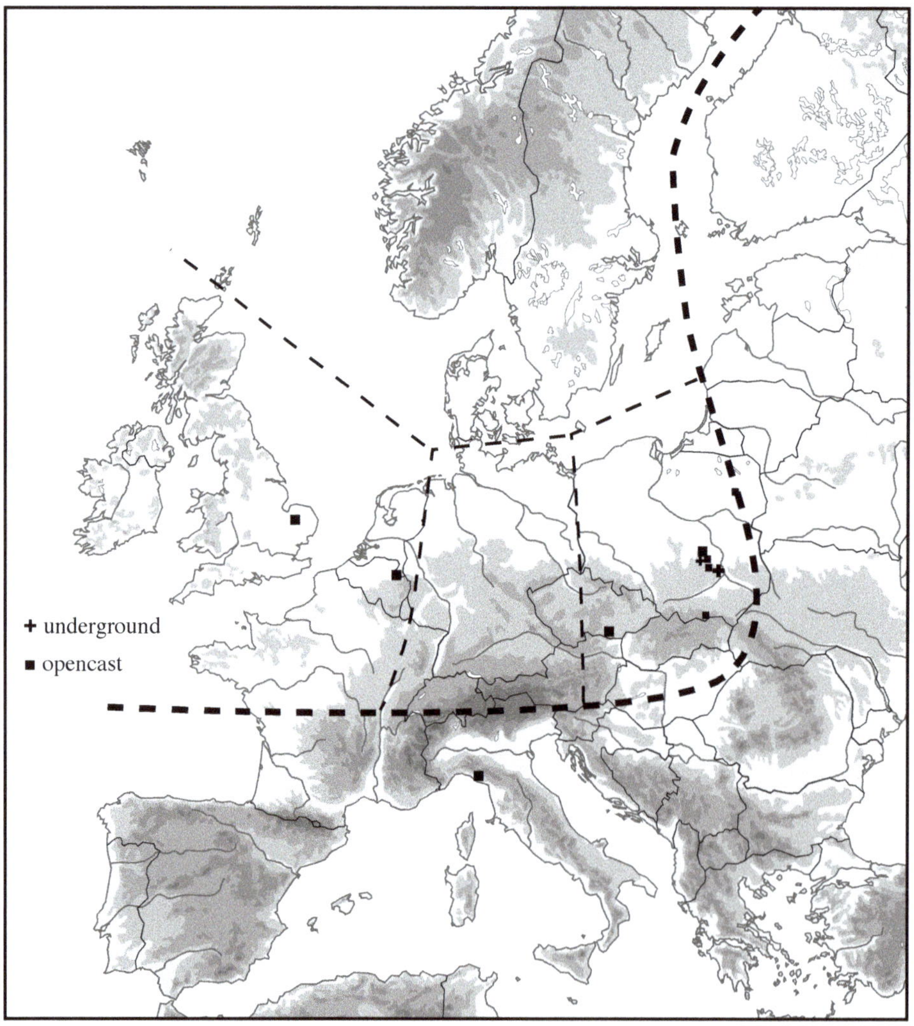

Fig. 8 : exploitation sites and traces of Metal Ages.

the Chalcolithic: complementarity, interdependence, extension over sometimes very large geographical areas, production and exchange of artefacts of a more complex nature than simply functional tools (Lech 1979: 45). These facts suggest intense and large-scale contacts already developed among Neolithic communities and cultural groups.

There was no real innovation as regarding extraction techniques probably employed during the Neolithic and those in use from the beginning of the Chalcolithic. Neolithic people knew where to find raw materials, and how to dig deep to extract them, as the water wells found in central Europe also attest. However, there was probably not the economic demand to exploit these raw materials as intensively as was done during the Chalcolithic. The new mining techniques appeared during this epoch are therefore linked to a growth of demand, not a change in its nature, which was similar to the previous era and obeyed the same basic rules (Lech 1991: 569).
Accordingly, we shall not argue for a *break* between Neolithic and Chalcolithic traditions of mineral exploitation, as could have happened in the case of a real technological innovation. This conclusion refutes the hypothesis that the origins of intensive mining techniques should be associated with the ore-mining occurring with the first metal production.

However, it would be wrong to conclude that there was a simple linear *evolution* between these two eras. At the beginning of the Chalcolithic, mineral exploitation not only displayed a new diversity and intensity, but also involved specialisation to a greater extent: communities now settled near the sources of minerals, in agriculturally poor areas previously uninhabited; the different phases of work were efficiently distributed in space and in accordance with individual abilities... Even though these phenomena may have been present to some extent during the Neolithic, during the Chalcolithic they took on so much importance that they generated a real 'mining engineering' expertise in some populations.

Thus we see a clear *transformation* in regard to mineral exploitation between the end of the Neolithic and the beginning of the Chalcolithic. Because of the extended, interconnected economic systems of the Neolithic, these changes happened almost simultaneously, during the early Chalcolithic, throughout the whole northern half of Europe. The main outcome was the expansion of the range of mining techniques; an intensive and technologically sophisticated production system with a capacity for deep mineral extraction spread quickly through the north-western margins of Europe (Scandinavia and British Islands). We also conclude that the appearance of new demands, involving new technological responses, coexisted with economic systems already established during the Neolithic and which persisted for some time.

2.2. Mineral exploitation during the Chalcolithic

During the early Chalcolithic, we observe the largest spectrum of mining techniques for the whole early early protohistory. The most extensive and elaborate techniques, such as deep quarrying of thick deposits and systematic underground networks, were generally adopted all over the northern half of Europe. Thus raw materials, for the large-scale production of standardised artefacts, mainly axes and blades, were extensively sought after, intensively exploited and sometimes widely diffused.

These products suggest a new notion of specialisation. During the early Chalcotlithic, the deepest and most intensive extraction techniques involved specialised populations, with considerable 'mining engineering' skills, acquired collectively and over generations of part-time miners. The *chaînes opératoires* of both extraction and production led, with some exceptions, to a generally efficient organization of standardised and easily communicable tasks, not requiring highly individual skills.

It is not easy to define the characteristics of these mining populations. Some evidence, such as the presence of abundant mining waste products in the surrounding settlements, suggests that they may have been established as village communities. However, the overall lack of compelling evidence incites us to characterise them with notions of *milieu technique*, or even *groupe technique*, as defined in the works of A. Leroi-Gourhan (1971: 321 ; 1973: 348). In any case, these populations did not necessarily carry on long-extended mining activities, but may have done so in repeated, regular and perhaps annual periods.

We shall therefore not compare these activities, as is sometimes done in archaeological literature (see for example Gardiner 1990: 122), to large-scale and infrequent collective works, such as the erection of megaliths or the construction of causewayed enclosure, which probably required the authority of an individual 'master-builder'. On the contrary, the collective experience involved in intensive mineral exploitation would have been progressively elaborated, through a repeated practice included in an 'ordinary' and probably annual economic organization.

The main tendencies of the early Chalcolithic economy persisted widely thereafter. During the middle Chalcolithic, new importance seems to have been given to the aesthetic properties of raw materials. This evolution, which may sometimes be linked to the production of prestige pieces with a clear symbolic value, did not directly involve changes in the nature and the scale of the mineral exploitation, but played an increasing role in the choice of the raw materials.

During the middle and late Chalcolithic, a progressive narrowing of the spectrum of exploitation techniques can be observed. Except in some notable cases, both the intensity and the complexity of these productions declined. The role of underground galleries clearly decreased; at the same time, the opencast and slope extractions, requiring a simpler know-how, gained in relative importance. Moreover, the products of mines exploited only for a short time, with rough and ready techniques, were not so much obscured by the predominance of standardised artefacts from major production sites as had earlier been the case.

During the late Chalcolithic, we observe almost exclusively simple opencast extraction techniques, sometimes linked with the production of at least partly non-functional items, such as the Danish and Swedish flint daggers and the long blades of Grand-Pressigny. These productions require high individual skills and experience, developed by each 'professional' knapper personally, contrasting with the collective nature of the know-how involved by intensive and complex mining techniques. This kind of 'professionalism', based on prestige craft production and appearing at the very end of the Chalcolithic, may constitute the first step towards the emergence of skilled artisans during the Bronze Age.

CONCLUSION

The conclusions driving from this study do not end discussions of this topic, far from it. Rather, they are meant to offer a general scheme of interpretation, which needs to be nuanced and no doubt corrected by new research into mineral exploitation, particularly through dialogue with the other disciplines concerned with lithic economies. In all events, these general observations are meant to show how the study of extraction techniques can contribute to defining and better understanding the main periods of early early protohistory, which really deserves to be more widely and more precisely studied

Yoann Gauvry
UMR 7041 ArScAn, équipe Protohistoire européenne
5 bvd des Alliés, 91720 MAISSE
e-mail: yoann.gauvry@wanadoo.fr

REFERENCES

Bąbel, J., Balcer, B., Chmielewska, M., Dagnan-Ginter, A., Ginter, A., Lech, J. and Schild, R. 1999. Katalog der Bergwerke – Poland. In Weisgerber G., Slotta R., Weiner J. (dir) : 5000 Jahre Feuersteinbergbau - Die Suche nach dem Steinzeit, Exposition au Deutschen Bergbau-Museum Bochum, du 24 oct. 1980 au 31 jan. 1981, éditions du Deutschen Bergbau-Museum Bochum, n°77, 3ème édition améliorée, élargie et actualisée, 578-627.

Balcer, B.1976. Position and stratigraphy of flint deposits, development of exploitation and importance of the Swieciechow flint in Prehistory. A.A.C. t. XVI, 179-199.

Becker, C. J. 1959. Flint mining in Neolithic Denmark. *Antiquity* 33, 130, 87-92.

Becker, C.J., Weisgerber, G. 1999. Katalog der Bergwerke – Dänemark. In Weisgerber G., Slotta R., Weiner J. (dir) : 5000 Jahre Feuersteinbergbau - Die Suche nach dem Steinzeit, Exposition au Deutschen Bergbau-Museum Bochum, du 24 oct. 1980 au 31 jan. 1981, éditions du Deutschen Bergbau-Museum Bochum, n°77, 3ème édition améliorée, élargie et actualisée, 456-473.

Biró, K.T. and Regenye, J. 2003. Exploitation regions and workshop complexes in the Bakony Mountains, Hungary. In Stöllner Th., Körlin G., Steffens G., Cierny J. (dir.) : Man and Mining - Mensch und Bergbau. Studies in honour of Gerd Weisgerber on occasion of his 65th birthday, Der Anschnitt Beiheft 16, Bochum, 55-63.

Borkowski, W. 1997. Exploitation field in Krzemionki. System of deposit utilization. In ' Siliceous Rocks and Culture ' Ramos-Millan et M.A. Bustillo Eds. VI International Flint Symposium, Madrid, 1-4 octobre 1991, Granada, 327-335.

Bostyn, F. and Lanchon, Y. 1992. *Jablines, le Haut Château (Seine et Marne); une minière de silex au Néolithique*. Document d'Archéologie Française 35. Paris: Maison des Sciences de l'Homme.

M. Capote, N. Castañeda, S. Consuegra, C. Criado, P. Díaz-del-Río ce volume. Flint mining in early Neolithic Iberia: a preliminary report on 'Casa Montero' (Madrid, Spain).

Clark, G. and Piggott, S. 1993. The age of the British flint mines, *Antiquity* 7, 166-183.

Collet, H. 2004. Les mines néolithiques de Spiennes : état des connaissances et perspectives de recherche, actes du XIV° congrès UISPP, section 9 : Le Néolithique au Proche Orient et en Europe, 129-133. *B.A.R. International Series* 1303, Oxford : BAR.

Collet, H., Collette, O. and Woodbury, M. 2004. Indices d'extraction et de taille du silex datant du Néolithique récent dans la carrière CBR à Harmignies. Note préliminaire. *Notae Praehistoricae* 24, *Studia Praehistorica Belgica*, 151-158.

De Labriffe, P.-A., Augereau, A., Sidéra, I. and Ferdouel, F. 1995. Villemaur-sur-Vanne 'Les Orlets' (Aube), quatrième et dernière minière de l'autoroute A5; résultats preliminaries. 19° colloque interrégional sur le Néolithique, Amiens 1992, *Revue Archéologique de Picardie*, n° spécial 9, 105-119.

Dijkman, W., Guillaume, Ch., Affolter, J. and Roden, Ch. 1999. Katalog der Bergwerke – Frankreich. In Weisgerber G., Slotta R., Weiner J. (dir) : 5000 Jahre Feuersteinbergbau - Die Suche nach dem Steinzeit, Exposition au Deutschen Bergbau-Museum Bochum, du 24 oct. 1980 au 31 jan. 1981, éditions du Deutschen Bergbau-Museum Bochum, n°77, 3ème édition améliorée, élargie et actualisée, 474-509.

Di Lierna, S., Fiorentino, G., Galiberti, A. and Basili, R. 1995. The early Neolithic mine of La Defensola 'A' (I18): flint exploitation in the Gargano area. In Lech J. (dir.) : Special Theme: Flint Mining, Actes du 7° symposium international du silex, Varsovie Ostrowiec Swietokrzyski 4-8 septembre 1995, Archaeologia Polona, vol. 33: 1995, Institute of Archaeology and Ethnology, Polish Academy of Sciences, Warsaw, 119-132.

Dzieduszycka-Machnikowa, A. and Lech, J. 1976. *Neolityczne zespoły pracowniane z kopalni krzemienia w Sąspowie* [The neolithic workshop assemblages from the flint mine of Sąspow], *Polskie Badania Archeologiczne* t. 19. Polska Akademia Nauk - Instytut Historii Kultury Materialnej.

Eisele, Kl., Rind, M. M. and Sorcan, B. 2003. Ausgrabungen und Dokumentationen im neolithischen Hornsteinbergwerk von Abensberg-Arnhofen 2000 bis 2002, *Rind*, 42-51.

Engel, F. and Siegmund, Fr. 2005. Radiocarbon dating of the Neolithic flint mine at Kleinkems (near Efringen-Kirchen, District Lörrach, Baden-Württemberg, Germany), *Antiquity* 79, 306. (website http://antiquity.ac.uk/projgall/siegmund/index.html).

Felder, P. J. 1999. Feuersteinbergbau in Ryckholt-St-Geertruid und Grime's Graves - ein Vergleich. In Weisgerber G., Slotta R., Weiner J. (dir): 5000 Jahre Feuersteinbergbau - Die Suche nach dem Steinzeit, Exposition au Deutschen Bergbau-Museum Bochum, du 24 oct. 1980 au 31 jan. 1981, éditions du Deutschen Bergbau-Museum Bochum, n°77, 3ème édition améliorée, élargie et actualisée, 120-123.

Felder, P. J., Rademakers, P., Cor, M. and De Grooth, M. E. Th. 1998. *Excavations of prehistoric flint mines at Rijckholt-St. Geertruid (Limburg, The Netherlands), by the 'Prehistoric Flint Mines Working Group' of the Dutch Geological Society, Limburg Section*, Archäologische

Berichte 12. Bonn: Deutschen Gesellshaft für Ur- und Frühgeschichte.

Galiberti, A., Di Lierna, S., Fiorentino, G. and Guarascio, M. 1997. New data on the Neolithic mine of Defensola, Vieste (Italy). In Ramos-Millan et M.A. Bustillo (eds). 'Siliceous Rocks and Culture' VI International Flint Symposium, Madrid, 1-4 octobre 1991, Granada, 211-222.

Gardiner, J. 1990. Flint procurement and Neolithic axe production on the South Downs: a re-assessment. *Oxford Journal of Archaeology* 9(2), 119-141.

Gayck, S. 2000. *Urgeschichtlicher Silexbergbau in Europa - Eine kritische Analyse zum gegenwärtigen Forschungstand.* Weissbach: Beiträge zur Ur- und Frühgeschichte Mitteleuropas 15.

Goldenberg, G., Maass, A., Steffens, G. and Steuer, H. 2003. Hematite mining during the Linear Ceramics Culture in the area of the Black Forest, South West Germany. In Stöllner Th., Körlin G., Steffens G., Cierny J. (dir.): Man and Mining - Mensch und Bergbau. Studies in honour of Gerd Weisgerber on occasion of his 65th birthday, Der Anschnitt Beiheft 16, Bochum, 179-186.

Gurina, N. N. 2000. *Prähistorische Feuersteinbergwerke in der ehemaligen UdSSR.* Der Anschnitt Beiheft 12. Bochum.

Lech J. 1979. Flint mining among the early farming communities of Poland. *Neerlandse Geologische Vereiniging,* 39-45.

Lech, J. 1981a. Flint mining among the early farming communities of central Europe, Przegląd Archeologiczny, (Wrocław) 28, 5-55.

Lech, J. 1981b. *Górnitwo krzemienia Społeczności wczesnorolniczych na Wyżynie Krakowskiej, koniec VI tysiąclecia – 1 połowa IV tysiąclecia p.n.e.* [Flint mining among the early farming communities in the Cracow upland (end of VIth millenium to mid-IVth millenium B.C.)], Polska Akademia Nauk - Instytut Historii Kultury Materialnej.

Lech, J. 1991. The Neolithic-Eneolithic transition in prehistoric mining and siliceous rock distribution. In Lichardus J. (dir.): Die Kupferzeit als historische Epoche, Symposium Saarbrücken und Otzenhausen 6.-13.11.1988, Saarbrücker Beiträge zur Altertumskunde, Band 55, Bonn, 2 vol., 557-574.

Leroi-Gourhan, A. 1971. *Evolution et techniques : l'homme et la matière.* Paris : Albin Michel.

Leroi-Gourhan, A.1973. *Evolution et techniques : milieu et techniques.* Paris: Albin Michel.

Lichardus, J., Lichardus-Itten, M., Bailloud, G. and Cauvin, J. 1985. *La Protohistoire de l'Europe, le Néolithique et le Chalcolithique.* Paris: Presses Universitaires de France, Nouvelle Clio.

Migal, W. 1997a. Selected aspects of specialization in mining and flint knapping. In Schild R., Sulgostowska Z. (eds) : Man and flint, actes du 7° symposium international du silex, Varsovie Ostrowiec Swietokrzyski 4-8 septembre 1995, Institute of Archaeology and Ethnology Polish Academy of Sciences, Warszawa, 99-101.

Migal, W. 1997b. Reconstruction of the flint extraction system in Krzemionki. In Ramos-Millan et M.A. Bustillo (eds). 'Siliceous Rocks and Culture' VI International Flint Symposium, Madrid, 1-4 octobre 1991, Granada, 315-325.

Pelegrin, J. 2002. La production des grandes lames de silex du Grand-Pressigny. In Guilaine J. (dir.) : Matériaux, productions, circulations du Néolithique à l'Age du Bronze, séminaire du Collège de France, Collection des Hespérides, Errance, Paris, 131-148.

Ruttkay, E. 1970. Dasjungsteinzeitliche Hornsteinbergwerk mit Bestattung von der Antonshöhe bei Mauer (Wien 23). Die Ausgrabungen Josef Bayers in den Jahren 1929-1930. *Mitteilungen der Anthropologischen Gesellschaft* 100, Wien, 70-83.

Shepherd, R., Sieveking, G. de G., Weisgerber, G. and Roden Ch. 1999. Katalog der Bergwerke – Grossbritannien. In Weisgerber G., Slotta R., Weiner J. (dir) : 5000 Jahre Feuersteinbergbau - Die Suche nach dem Steinzeit, Exposition au Deutschen Bergbau-Museum Bochum, du 24 oct. 1980 au 31 jan. 1981, éditions du Deutschen Bergbau-Museum Bochum, n°77, 3ème édition améliorée, élargie et actualisée, 510-543.

Schmid, E.1952. Jungsteinzeitliches Jaspis-Bergwerk am Isteiner Klotz. *Der Anschnitt* 5/1952, 4. 4-14. Bochum.

Schmid, E.1973. Die Reviere urgeschichtlichen Silexbergbaus in Europa. *Der Anschnitt* 4/1973, 25, 12-15. Bochum.

Schmid, E.1974. Die Reviere urgeschichtlichen Silexbergbaus in Europa - Teil IV: Die Silex-gewinnung in Nordfrankreich. *Der Anschnitt* 3/1974, 18-22. Bochum.

Weiner, J. and Weisgerber, G. 1999. Die Ausgrabungen des jungsteinzeitlichen Feuersteinbergwerks 'Lousbergww in Aachen 1978-1980 (D3), Ein Vorbericht. In Weisgerber G., Slotta R., Weiner J. (dir) : 5000 Jahre Feuersteinbergbau - Die Suche nach dem Steinzeit, Exposition au Deutschen Bergbau-Museum Bochum, du 24 oct. 1980 au 31 jan. 1981, éditions du Deutschen Bergbau-Museum Bochum, n°77, 3ème édition améliorée, élargie et actualisée, 92-119.

IDEOLOGY AND INFLUENCES BEHIND THE NEOLITHIC FLINT MINES OF THE SOUTHERN BRITAIN

Paul Wheeler

Abstract: This paper presents the results of part of a research project which aims to evaluate the influences and evolution of the early Neolithic in Southern Britain. The following discourse endeavours to understand the development of the early Neolithic flint mining tradition, with particular reference to the Sussex mines of southern Britain.
The chronologically contemporary flint mines of Western Europe are analysed, in order to determine if common practices occur amongst the mines of Europe and those that first appeared in Britain and thus uncover any potential continental influences on the early Neolithic of Southern Britain. The research looks beyond superficial similarities, and looks to identify commonalities in the ideology and the attitude associated with the mines and the mining process. Do we find similar events taking place in the mines? Are technical problems dealt with in similar ways? and how does the approach towards flint manifest itself? The common practices identified are then analysed in order to determine if an ideological bond can be found throughout the flint mining process of Western Europe. Ideas are proposed about what we can infer from communities who are displaying a similar set of cultural choices across this part of Europe.

Keywords: Flint mine, western Europe, Neolithic.

INTRODUCTION

In Southern Britain one of the first truly monumental structures to be created within the early Neolithic were the flint mines of the Sussex downs. The landscape begins to be peppered with these deeply intrusive, exotic and complex subterranean structures, and so begins one of the earliest truly monumental endeavours of the local communities of Britain. The technological and ideological concepts behind the creation of the mines would presumably have been a revolutionary concept, with far reaching implications on the politics and economics of the societies involved.

Thus the initial appearance of the flint mines prompts two observations. Firstly, a change in the mindset and mentality of the populous must have taken place. Previously local communities would have collected flint from river beds or from outcrops (Barber 2001: 22-23), however, now local communities had the desire to obtain flint from deep within the landscape. Why did this change take place? Perhaps there was a need for a greater quantity or purer quality of flint. Or perhaps an altogether different ideology and attitude towards the procurement of flint was created. The question thus arises as to whether the idea and ability to mine flint was an indigenous innovation or could such concepts have been introduced from the continent.
Thus the main objective regarding the analysis of flint mines is to critically evaluate the evidence for a continental influence for their establishment in the British landscape in the early Neolithic.
In order to establish any potential influences on the development of the mines, this paper looks beyond a morphological and technological analysis, upon which variation and commonalities can be explained via environmental determinism, and looks at the cultural choices undertaken by each community. If there was a continental influence upon

the flint mines of Sussex, one would expect to identify a series of common cultural practices, which are inherently associated within flint mining traditions.

1. BACKGROUND TO THE RESEARCH

There where three main criteria for the selection of comparative flint mine sites throughout Europe.

Firstly the sites had to fall into a chronologically contemporary period as the Sussex Mines. Thus this study will examine the sites that were in operation in the last few centuries of the 5th millennium BC and for the first half of the 4th millennium BC. The chronology for the mine sites are based on the earliest recorded calibrated dates.

Secondly the mines had to lie within the pre determined study region, which consists of Britain, Belgium, France, The Netherlands and western Germany (Fig. 1). Thirdly, only mines containing a large enough excavated and recorded sample range were analysed, in order to have a justifiably comparable dataset. The British mines under study consist of the Sussex mines of Church Hill (Barber *et al.* 1999; Russell 2001: 85-158), Harrow Hill (Curwen and Curwen 1926; Holleyman 1937), Cissbury (Russell 2001: 170-192; Topping 2005: 63-93) and Blackpatch (Pull 1932; Russell 2001: 24-84).

Across the channel we find the French mines of Bretteville-le-Rabet (Desloges 1986) and Jablines (Bostyn and Lanchon 1992 ; 1997) are being exploited within a similar period, as are the Belgic mines of Jandrain Jandrenouille (Hubert 1974; Weigsberger 1980: 429-431) and the mines at Spiennes (Hubert 1978; Weigsberger *et al.* 1980: 418-421). The Dutch mine of Rijckholt-St-Geertruid (De Grooth 1998; Bosch 1979; Felder *et al.* 1998; Felder 1981) is also in use during a this time (Fig. 2).

The chronology of the sites is based on the latest and earliest recorded calibrated dates obtained for each site under study. The date ranges indicate the mines were in operation at approximately the same time, the data indicates an average common exploitation period of approximately 3800BC (Fig. 2).

2. WHAT ARE THE COMMON PRACTICES?

A number of common cultural choices can be explored, the concept of mining for flint, the flint seam exploited, the products manufactured at the mine and finally an analysis of the various types of deposits within mine shafts. The first commonality between the Sussex mines and those on the European continent is the construction and concept of flint mining itself.

The flint mines themselves are the construct of a totally

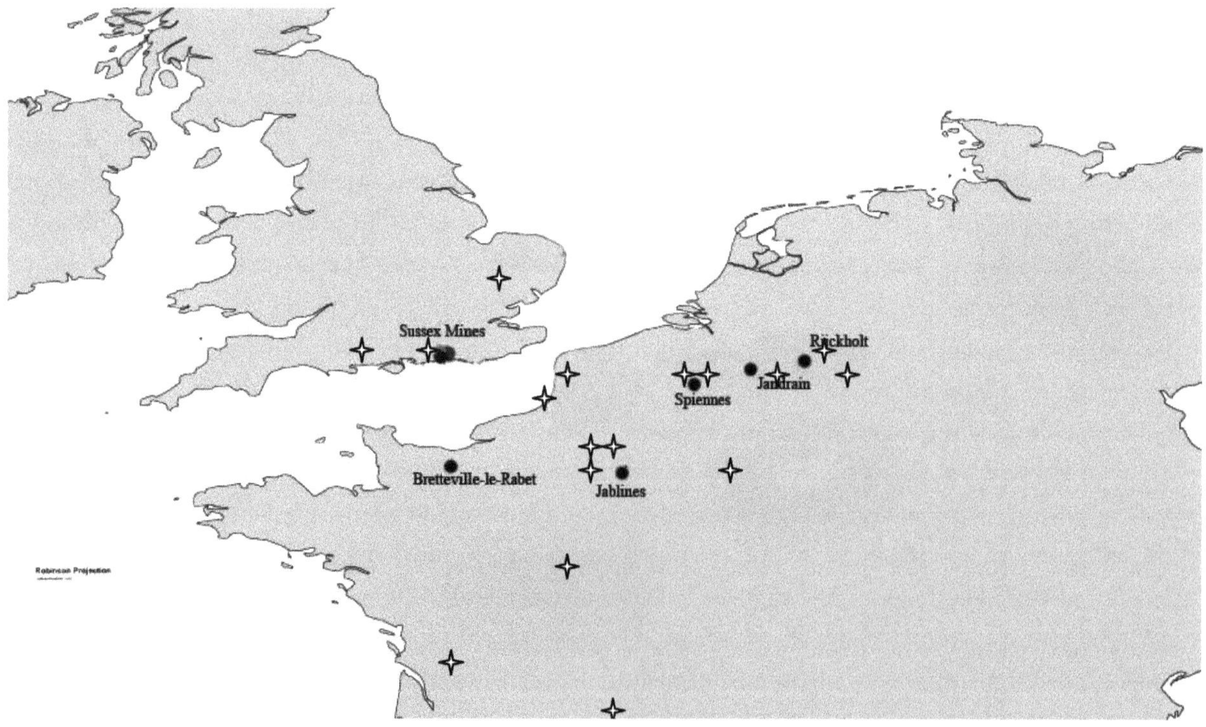

Fig. 1: Western European Flint mining sites and those under study.

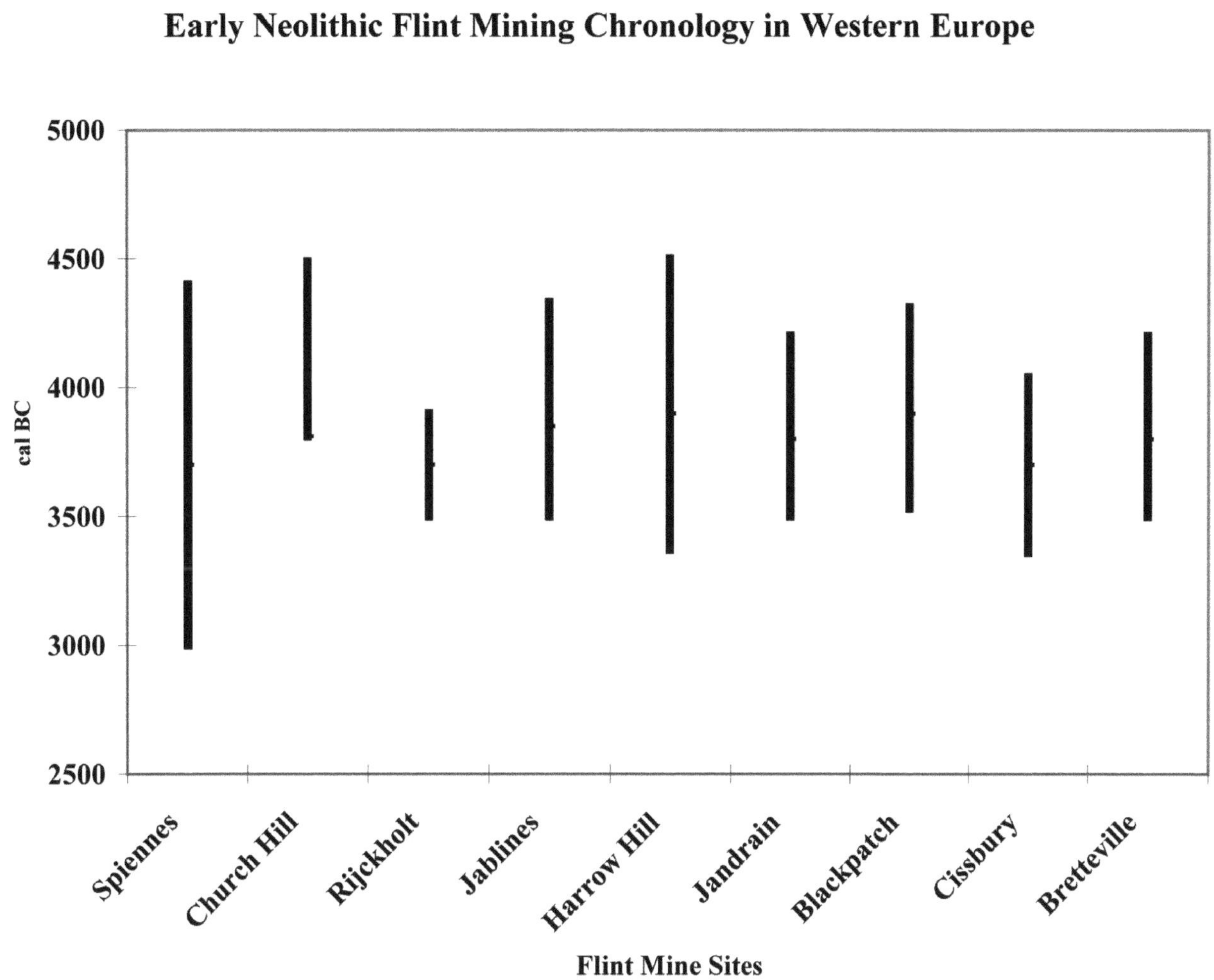

Fig. 2: Early Neolithic Flint Mining chronology for Western Europe.

alien technological ability. Nothing on this scale had previously been created within southern Britain. Thus the initial occurrence of flint mines equates not only to a new concept but a new technological skill amongst the local communities who now not only posses the engineering ability to construct these great subterranean monuments but also the desire to intrusively exploit the earth for this raw material. But how did the indigenous populations obtain such skills and evolve the desire to mine for flint?

Looking across to the European mainland we find that the mining communities at Spiennes, Rijckholt and those other mines in the study region, are harnessing and exploiting similar methods in the extraction of flint (Felder *et al.* 1998; Hubert 1978). Whether in the form of small pits, quarries or big, deep shafts and subterranean galleries, the concept of digging deep into the earth is abundantly evident, as are the technological and engineering skills used to sustain such a vast monumental complex (Pull 1932: 26).

Thus the overall concept and cultural choice to mine for flint can be identified as a common practice across the continent and thus equate to potential cultural influence.

3. THE CHOICE OF FLINT SEAM TO BE EXPLOITED

The process of flint mining is largely determined by environmental factors. Firstly, you can only mine in a geological landscape that contains flint and of a reasonable quality, so communities had little choice over location.

Secondly the size of a shaft is determined by the structural integrity of the underlying geology. This also affects the tools used to dig shafts.

Thus the first quantifiable culturally determined choice, that miners appear to have made is, which flint seam should be exploited?

Examination of the shafts evaluated to date have identified that the majority of the miners throughout this period in

Europe were not exploiting the first or second seams of flint they encountered during the process of extraction, but were exploiting the deeper seams (Fig. 4), often bypassing the more easily accessible flint above (Curwen and Curwen 1926; Felder *et al.* 1998: 39; Pull 1932: 25).

The evidence from the Sussex Downs shows that at Church Hill the first 3 seams of flint were by-passed before exploitation took place (Russell 2001: 85-158). At Cissbury the 3rd and 4th seams were those exploited (Russell 2001: 179-192), and at Harrow Hill it was mainly the 3rd seam to be exploited, with the first seams dug straight through (Fig. 3) (Curwen and Curwen, 1926).

Only at Blackpatch was the first seam encountered exploited, but even here, as amongst the other Sussex shafts, miners would dig down up to 15ft to access the desired material (Pull 1932: 25). Such cultural choices of the Sussex miners are witnessed amongst the mines of the European mainland. At Spiennes the miners would often dig through numerous seams to exploit the lowest safely accessible seams, reaching depths of up to 16m (Fig. 4). The flint at this level was up to 30cm thick, quite different from the seams above (Hubert 1978). The miners at Rijckholt would also by-pass the first easily exploitable seams, looking to utilise the specific lanaye band of flint (Fig. 3), which commonly occurred at seam 10 (Felder *et al.* 1998, 12).

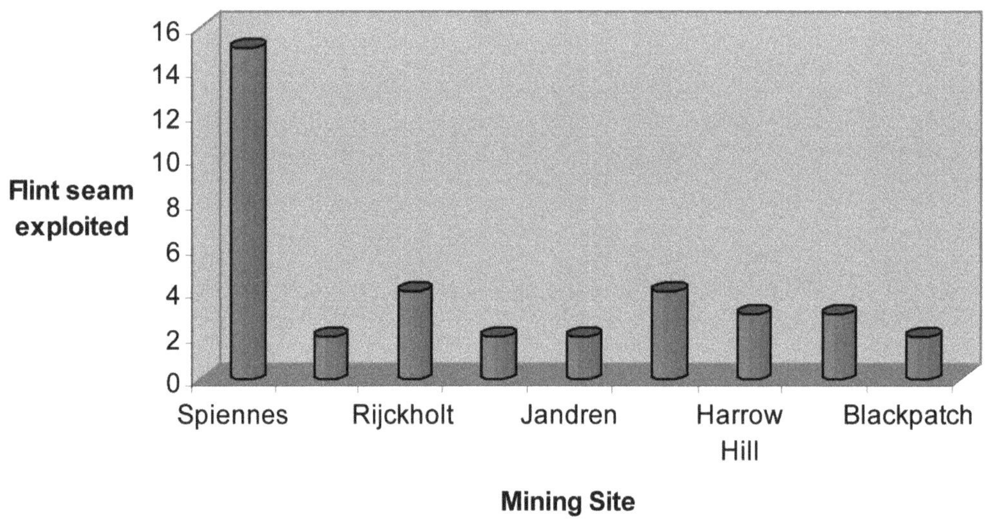

Fig. 3: Average flint seam exploited in the mine sites under study.

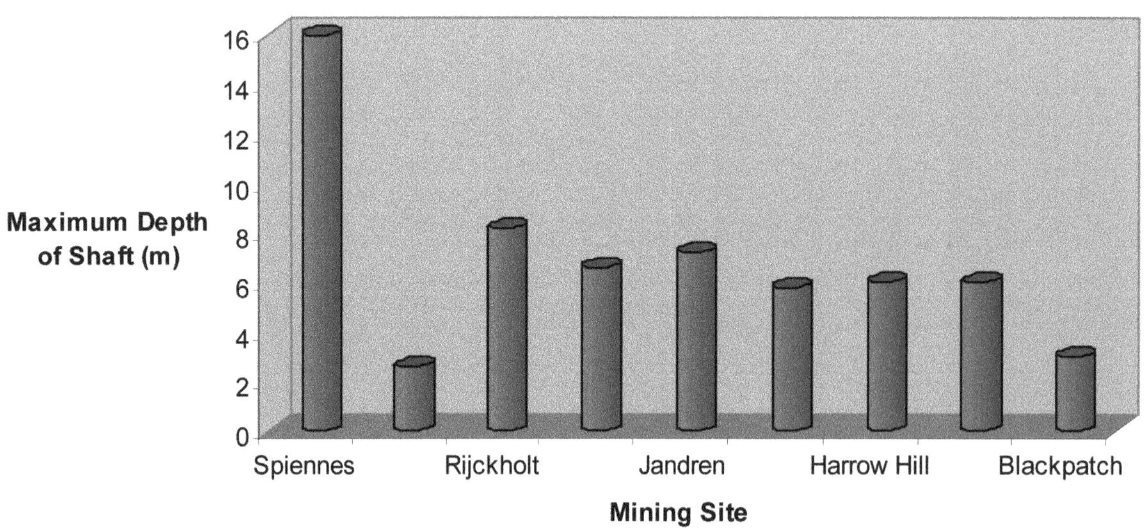

Fig. 4: Maximum depht of shaft within the mine sites of under study.

At Jandrain-Jandrenouille and Jablines, although 1st and 2nd seams were exploited (Fig. 3), the required flint continued to be difficult to access; often these mines reached depths of 5m and 6m respectively (Hubert 1974; Bostyn and Lanchon 1992: 58-87).

The evidence from across Europe shows that mining communities were making a cultural choice to either exploit particular seams of flint, or they were choosing to exploit seams which were more difficult and dangerous to access (Barber 2001: 23).

Figure 4 shows the comparative depths the deepest mines at each site would reach. But what determined this choice? In regards to cases where specific seams are exploited, it appears that implicit values may be attached to these particular seams.

Where miners are choosing to dig deep into the earth, and bypassing other exploitable seams, we may be witnessing episodes on a par with those witnessed amongst stone axe procurement sites, where it is the difficulty of obtaining the raw material that creates the social values imbedded within the objects of production (Topping 2005: 83-84).

However, it could be that certain flint seams were just not physically suitable for the production of the objects to be manufactured and were thus rejected for exploitation.

The evidence from across Europe shows that although there was plenty of surface flint, easily and readily available, communities within the early Neolithic were prepared to go to great lengths to access raw material from deep within the earth (Barber 2001: 22-23). Even the unexploited seams and poorer quality seams within the mine shafts could have been utilised for the large scale production of routine blades, scrapers and knives, but were barely utilised. Why? Perhaps it is the association of the mine shafts and the items produced there.

4. THE CHOICE OF PRODUCT TO BE MANUFACTURED

The data from the mines within the study region indicates that a commonality occurs between the objects of production and the procedures involved in their manufacture.

Firstly the mines of Sussex, Jablines, Bretteville and Jandrain are all mainly manufacturing axes (Russell 2001; Bostyn and Lanchon 1992; Desloges 1986; Hubert 1974). At the larger mines of Spiennes and Rijckholt, again we find axe manufacture integrally associated with these mines, however blade manufacture appears to be an important part in their later phases (Hubert 1978; Weigsberger 1980: 418-421; De Grooth 1998; Bosch 1979; Felder *et al.* 1998; Felder 1981).

Axe manufacture appears to be integrally linked to flint mining, particularly in the early Neolithic (Topping 1997). All the mines have a strong association with the production of flint axes. If mine shafts are only to be used for the production of axes this may explain some of the choices identified so far. Perhaps it is this association that determines which flint seams are exploited?

Firstly perhaps only particular seams are of a fine enough quality for the production of axes, for example the Lanaye layer at Rijckholt, the thick seams at Spiennes or even the utilised layer at Church Hill. But why were the lesser quality seams not also exploited for more commonplace products? Perhaps the mine shafts were solely the prestige of axes, and thus the lesser seams could not be utilised for the production of secondary (less prestigious) products.

Thus perhaps axe manufacture holds an exclusive status amongst the flint mines. The depths miners were prepared to dig to access particular seams and the inherent dangers involved may have added an implicit value to the axes. The questions and values associated with ëdifficulties of accessí, can be mirrored amongst stone axe production sites.

Barber, 2001 has suggested regarding flint mines and quarry sites, that perhaps it is the location and manner of exploitation that highlights the desire to obtain stone in a particular way or from particular place.

The importance of the axe and its inherent association with flint mines is a common thread that is shared between the mines across Europe. This commonality is further supported by the use of similar procedures in regard to the manufacture of axes at these sites. The process of manufacture is always the same throughout all the sites in the study region. The mined raw flint is chipped to form roughouts at the site before being transported to secondary sites for final polishing and finishing (Bostyn and Lanchon 1992: 136-139; Desloges 1986; Russell 2001: 45-48). There is very little evidence for polishing or final polished products at the mines (Pull 1932: 112-114; Collet *et al.* 1997; Desloges 1986).

Yet again we find similar echoes within the chronologically contemporary axe factories at Great Langdale, Cumbria (Bradley and Edmonds 1993: 37-38) and Seledin, France (Le Roux, 1971), where not only do we find similar issues regarding difficulty of access, but the manufacturing procedure also follows similar patterns with the axes being formed to roughout on the site and transported to secondary polishing sites.

The parallels in not only the products produced, but also the method of manufacture across the mine sites throughout Europe, continues to imply a shared influence and ideology associated with the process of mining and perhaps more importantly axe manufacture.

5. THE CHOICE OF DEPOSITION

The final cultural choice associated with flint mining, explores how the excavated mine shafts were treated during and after the extraction process. Almost all the mines have evidence of depositional activities. However, are these deposits mere random acts of waste disposal or can we identify evidence of set predetermined behaviours amongst the mining communities across Europe to the mines themselves.

Firstly we will explore some of the evidence for deliberate depositional events within certain shafts throughout Europe; these examples consist of certain selected deposits within shafts, which would appear to be beyond identification as a waste disposal deposit.

The most documented evidence exploring the idea of attitudes towards the shafts, is presented by Peter Topping (2005), who integrated an ethnographical framework into hypothesised formal structured depositional events with the backfill of shaft 27 at Cissbury, Sussex. The first episodes of deposition within the shaft were associated with cleansing rituals and offerings; subsequent deposits were associated with the process of extraction while later episodes of deposition are associated with tool production and rites of renewal (Topping 2005: 78-79). The near complete skeleton of an ox at 3.5m can be interpreted as an offering, as it is an almost complete ox, while it could also be a waste deposit or an unfortunate accident. Further deliberate deposits within this shaft are the numerous axes and axe roughouts and the deposition of the human female skeleton lying at the entrance to the west gallery with what appear to be two chalk charms in her right hand (Topping 2005: 71-72).

Similar episodes of what appear to be deliberate depositional episodes can be identified throughout the other mines of Sussex. For example, within the backfill of shaft 7 at Blackpatch, we find a cremation deposit associated with a flaked axe, scraper, knife and 'a curious charm of worked chalk' (Pull 1932: 56-58). Further down the shaft fill, lay another deposit consisting of animal bone, flint working debris and a pile of unused flint nodules, which according to the excavator were placed there (Pull 1932: 49). Within shaft 4 the disarticulated remains of an adult human femur and a jaw bone from a young person were identified. The details of these deposits appear to have been placed within the shafts, after the excavation of the raw flint material, for a purpose.

Purposeful deposits can be identified within the backfill of shaft 4 at Church Hill, which consists of 15 episodic events both natural and deliberate. Within these, deposits of fine, flaked axes are identified along with various flint working floors.

Also identified within the shaft were a series of unique geometric impressions above the gallery entrances, which cannot be paralled amongst any other mine sites (Russell 2001: 87-103).

At Harrow Hill, amongst shaft 21 and its various galleries similar deliberatedepositional practices can be identified. Antler picks, some in an unused condition were identified at the entrance to galleries (Curwen and Curwen 1926). Towards the base of the shaft a broken axe was deposited, with the 2 pieces placed 45cm apart, with different patination (Holleyman 1937: 239-242).

Such practices associated with the backfilling of the shafts and deliberate deposition in the galleries can be witnessed amongst the flint mines in Western Europe.

At Jandrain we find deposits of virgin unused picks and roughout axes within the galleries (Hubert 1974: 15). Similar practices are recognised at Bretteville-le-Rabet, deposits of unused tools are also found within the shafts. Further evidence from the shafts suggests tools were deposited at the base of exploited seams (Desloges 1986: 95). At Jablines, Rijckholt and Spiennes, large numbers of roughout axes have been identified deposited within shafts and galleries (Bostyn and Lanchon 1992; Felder *et al.* 1998; Verheyleweghen 1966). Further evidence of deliberate deposition can be identified when considering deposits of human remains. At Rijckholt firstly, a skull with no mandible was deposited at the end of one of the galleries, in a shallow pit

(De Grooth 1998). The gallery was later backfilled with chalk, except within the area of the skull (Bosch 1979; De Grooth 1998). Very similar practices have been identified at Spiennes. Buried skulls have been identified without the mandibles, these deposits have previously been associated with a specially protected phase (Verheyleweghen 1966).
The human remains identified in the Sussex mines however do not echo identical practices. At Blackpatch the disarticulated remains of an adult femur and human jaw bone from a young person were identified within shaft 4 (Pull 1932: 49), while as previously described, at Cissbury the full skeleton of an adult woman was found at the entrance to a gallery in shaft 27 (Russell 2001: 178-184).
The evidence indicates that throughout Europe, the mining communities chose to make certain deposits within the shafts, most commonly after mining and during the backfill process. Such deposits consist of either human remains placed in a particular way, as at Rijckholt and Spiennes, or often they consist of deposits of stone, axes or tools. However there does not seem to be a universal ideology, in the individual deposits. One would not have expected to find a regimented practice of deposition throughout Europe, and within each mine and individual shaft, but we can perhaps see a general trend and attitude to the mines in which the individual interpretation of how to represent a 'general' belief or attitude is flexible and dynamic. We may be seeing a general reflection of the choice and desire to give back to the earth, perhaps echoing hypotheses regarding rights of renewal.

6. DISCUSSION

The analysis of the cultural choices taken place within the flint mining process 'supports the evidence for common cultural ideology in respect to the attitudes and behaviour towards the mining process' and the mines themselves. Common practices have been identified throughout Europe in regards to the desire to mine for particular flint seams with implicit values and to produce similar products in similar ways. This analysis has also highlighted the choices made by cultures throughout Europe to make certain deposits within the shafts.
However it must be emphasised that this study is looking at broad generalisations and ideologies. As has been identified throughout the depositional analysis, there are differences of interpretation not only on a micro regional level but within individual mine site, from shaft to shaft. Individual shafts within a mine site would not only have been dug by different individuals, but also differing generations, kinships and perhaps cultures. Each individual could theoretically choose to represent their belief in different ways and thus we find a variety of manifestations and interpretations but all adhering to a general philosophy and ideology.
The evidence implies that mining and quarrying were largely for the construction of axes. There may have been an overall ideology that when this process was taking place either certain seams of flint or the difficulty of procurement, would add prestige and a cultural value to the object of production. However each culture or individual would choose at which point during the mining process they believed they had satisfied this obligation.
The second general ideology identified across Europe is one of 'giving back to the earth', where once extraction had been completed, the individual was obliged to create a deposit, an offering so to say. Again this offering could follow many different manifestations, whether in the form of human bone, axes themselves or tools, each individual would choose how they would satisfy there obligation under this general cultural ideology. Thus the range of deposits can be different from site to site and shaft to shaft across Europe. In general though we find that both cultures in Sussex and across Western Europe are following this general cultural ideology.

CONCLUSION

The mining communities across Europe each employ a similar set of cultural choices in their attitudes to flint mining. These common practices first show that across Western Europe, cultural ideas not just in the physical manifestation of mining technology and techniques are being transferred, but perhaps the ideology and beliefs integrated with such activities and process are being adopted as well. The action of extracting stone from the earth for the creation of axes appears to be integral to this common ideology, which is emphasised by cultural choices and the use of structured depositional practices.
The discussion above outlines the possibility of a general philosophy and ideology that underpins and unites Neolithic communities across Europe. It is stressed however that this encompasses general trends across a very large

area. The dynamic nature of beliefs themselves added temporal and spatial variability explains the differences in the micro detail between regions, mine sites and shafts.
Ultimately however, the cultural choices identified amongst the mines of Europe are being practiced within the Sussex mining tradition, implying that a continental influence may lie behind their origin, development and philosophy of use.

Paul Wheeler
8 Damask close, Tring, Herts, HP23 5UA.
e-mail: pabs42@hotmail.com

REFERENCES

Barber, M. 2001. Flint Mines in the early Neolithic and beyond. Raw material sources and later prehistory in Southern England. In A.T.Smith and A. Brooks (eds) Holy Ground: Theoretical Issues Relating to the Landscape and Material Culture of Ritual Space. Papers from a session held at the Theoretical Archaeology Group conference, Cardiff 1999. Oxford. BAR International Series 956, 21-26.

Barber, M., Field, D. and Topping, P. 1999. The Neolithic Flint Mines of England. Swindon. London. Royal Commission on the Historical Monuments of England/English Heritage.

Bosch, P.W. 1979. A Neolithic Flint Mine. Scientific American 240 (6):126-132

Bostyn, L. and Lanchon, Y. (eds) 1992. Jablines le Haut Château (Seine-et-Marne): une minière de silex au Néolithique. Paris. Maison des Sciences de l'homme. 246

Bostyn, L. and Lanchon, Y. 1997. The neolithic flint mine at Jablines 'Le Haut-Château' (Seine et Marne). In 'Siliceous Rocks and Culture' Ramos-Millan et M.A. Bustillo Eds. Actes du VI International Flint Symposium, Madrid, 1-4 octobre 1991, Université de Granada, p. 271-292.

Bradley, R. and Edmonds, M. 1993. Interpreting the axe trade: Production and exchange in Neolithic Britain. Cambridge. Cambridge University Press.

Collet, H., Deramaix, I., Sartieaux, P-P. and Vander Linden, M. 1997. Fouille préventive de puits d'extraction de silex à Petit-Spiennes (Hainaut). Notae Prehistoticae, 17, 203-212.

Curwen, E. and Curwen, E.C. 1926. Harrow Hill flint mine excavations 1924-5. Sussex Archaeological collections, 67, 103-138.

De Grooth, M.E.Th. 1998. The Flint Mines at Rijckholt-Sint Geertruid and Their Socio-economic Interpretation. In M.Edmonds and C.Richards (eds) Understanding the Neolithic of North-Western Europe. Cruithne Press. Glasgow, 351-369.

Desloges, J. 1986. Fouilles de mines à silex sur le site Néolithique de Bretteville-le-Rabet (Calvados). Revue archéologique de l'Ouest, supplement no 1, Rennes, 73-101.

Felder, P.J. 1981. Prehistoric flint mining at Rijckholt ñ St ñ Geertruid (Netherlands) and Grimes Graves (England). In F.H.G Engeln (ed) Proceedings of the Third International Symposium on Flint, 24-27 mei 1979 ñ Maastricht. Staringia 6, 1981, 25-28.

Felder, P.J., Rademakers, P. Cor M., and De Grooth, M.E.Th. 1998. Excavations of Prehistoric flint mines at Rijckholt-St.Geertruid (Limburg, The Netherlands) by the 'Prehistoric Flint Mines Working Group' of the Dutch Geological Society, Limburg Section. Bonn. Archaologische Berichte 12.

Holleyman, G. 1937. Harrow Hill Excavations, 1936. Sussex Archaeological Collections, 78, 230-251.

Hubert, F. 1974. Minières néolithiques à Jandrain-Jandrenouille en Brabant. Archaeologica Belgica, 167, 5-44.

Hubert, F. 1978. Une minière néolithique à silex au Camp-à-cayaux de Spiennes, Archaeologica Belgica, 210, 1-43

Le Roux, C.T. 1971. A stone axe-factory in Brittany. Antiquity, 45, 283-288.

Pull, J.H. 1932. The Flint Miners of Blackpatch. London. Williams and Norgate.

Russell, M. 2001. Rough quarries, rocks and hills : John Pull and the Neolithic flint mines of Sussex. Oxford. Oxbow.

Topping, P. 1997. Structured deposition, symbolism and the English Flint mines. In R.Schild and Z.Sulogostowska (eds), Man and Flint: Proceedings of the VIIth International Flint Symbosium, Warszawa ñ Ostrowiec Swietokrzyski, September 1995. Warszawa. Institute of Archaeology and Ethnography, Polish Academy of Sciences, 127-132.

Topping, P. 2005. Shaft 27 Revisited: An Ethnography of Neolithic Flint Extraction, In Topping, P. and Lynott, M. (eds) The Cultural Landscape of Neolithic Mines. Oxford. Oxbow, 63-93.

Verheyleweghen, J. 1966. Le Néolithique minier belge: son origine et ses relations culturelles. Palaeohistoria, 12, 529-557.

Weigsberger, G., Slotta, R. and Weiner, J. (eds) 1980. 5000 Jahre Feuersteinbergbau. Die Suche nach dem Stahl der Steinzeit. Verofentlichungen aus dem Deutschen Bergbau Museum Bochum 22.

Whittle, A. 1995. Gifts from the earth: symbolic dimensions of the use and production of Neolithic flint and stone axes. Archaeologia Polona, 33, 247-260.

ACKNOWLEDGMENTS

The writer would like to thank François Giligny for the kind invitation to speak at the EAA session in Cracow and also to thank Pierre Allard, Françoise Bostyn and Jacek Lech for organizing the session. Further thanks to Alan Saville, Yoann Gauvry, Anne Hauzeur and Marjorie de Grooth for all their comments and assistance in contributing and developing ideas to the research. James Morris also kindly read and commented upon the early draft of the paper.

www.ingramcontent.com/pod-product-compliance
Lightning Source LLC
LaVergne TN
LVHW071214210926
848482LV00015B/808